TERTIARY LEVEL BIOLOGY

Microbial Energetics

EDWIN A. DAWES, Ph.D., D.Sc., C.Chem., FRSC
Reckitt Professor and Head
Department of Biochemistry
University of Hull

Blackie

Glasgow and London
Distributed in the USA by
Chapman and Hall
New York

Blackie & Son Limited,
Bishopbriggs, Glasgow G64 2NZ

Furnival House, 14–18 High Holborn, London WC1V 6BX

Distributed in the USA by
Chapman and Hall
in association with Methuen, Inc.
29 West 35th St., New York, NY 10001
© 1986 Blackie & Son Ltd
First published 1986

British Library Cataloguing in Publication Data

Dawes, Edwin A.
 Microbial energetics.—(Tertiary level biology)
 1. Microbial metabolism
 I. Title   II. Series
 576'.1133   QR88

 ISBN 0-216-91790-5
 ISBN 0-216-91791-3 Pbk

For the USA, International Standard Book Numbers are
 0-412-01041-0
 0-412-01051-8 (Pbk)

Photosetting by Thomson Press (India) Limited, New Delhi
Printed in Great Britain by Bell & Bain (Glasgow) Ltd

Preface

This book is aimed at the advanced undergraduate in biochemistry and/or microbiology, and at the research worker who is entering the field and requires a succinct account of the various aspects of microbial energetics. By bringing together the broad spectrum of microbial energetics in a concise but readable form, properly cross-referenced to emphasize the inter-relationships of different aspects of the subject, I have endeavoured to provide the reader with a balanced survey within a restricted number of pages. A reference section is included at the end of the book, to allow particular points to be followed up in detail.

I am grateful to two of my colleagues: Dr George W. Crosbie for his constructive criticisms of the draft, and Dr Peter J. Large for his helpful comments on methanogens and methylotrophs; to Jane Naylor, Susan Wheeldon, Anne McLeish and Julie Foottit for their efficient typing of the manuscript; and to Carol Gillyon, Jennifer Curzon and Derek Waite for preparing the diagrams.

EAD

Contents

CHAPTER ONE

INTRODUCTION

Micro-organisms have always held tremendous fascination for biologists. Primitive microbes were the earliest forms of life on this planet, and today their successors inhabit regions as diverse as the soil, polar ice caps, thermal springs, deep oceans, salt lakes, sulphur springs, the gastro-intestinal tracts of animals, and polluted environments of virtually every kind. Their metabolic versatility, encompassing abilities to grow under aerobic or anaerobic conditions, with simple or complex nutritional requirements, has long attracted the attention of the biochemist, and microbes have frequently served as model systems for studies of the fundamental unity of life at the cellular level. The ability of micro-organisms successfully to colonize such contrasting habitats largely reflects their capacities to utilize the physical and/or chemical energy presented by these different milieux.

This book aims to provide a general survey of the energetics of micro-organisms and assumes that the reader will possess some basic knowledge of biochemistry and microbiology. It considers the processes whereby radiant energy and the free energy of chemical reactions are transduced for the growth and maintenance of the microbial cell under conditions which prevail in the natural environment. In the next section of this introduction an overview of these topics is provided and, to help the reader to locate the relevant chapters of the book, cross-references are given.

1.1 Synopsis of the book

Micro-organisms, in common with all other living organisms, require a continuous supply of energy for their growth and maintenance. This need may be met either by radiant energy or the free energy of chemical reactions, in the case of phototrophs or chemotrophs respectively, the potentially

1

available energy from both sources being transduced during the synthesis of ATP which acts as the link between sources of free energy and the biosynthetic reactions and other processes of the cell requiring an input of free energy. The role of ATP in metabolism is considered in Chapter 2.

Bacterial photosynthesis differs from that of green plants in being anoxygenic, i.e. oxygen is not evolved. The discussion of microbial photosynthesis in Chapter 10 therefore draws comparisons between the process of photophosphorylation in bacteria and that occurring in algae and cyanobacteria which resemble green plants in possessing an oxygenic system.

Chemotrophs secure their energy by the metabolism of appropriate substrates, usually organic compounds as exemplified by the chemoheterotrophs (Chapter 3), in processes which may be either (1) aerobic, when oxygen serves as the terminal electron acceptor in the oxidation of the compound to carbon dioxide and water, or (2) anaerobic, when either other organic compounds must serve as electron acceptors and the process is referred to as fermentation, or nitrate or sulphate function as electron acceptors in anaerobic respiration. Aerobic oxidation displays greater efficiency than anaerobic fermentation in terms of the ATP yield per mole of substrate metabolized. Under aerobic conditions the majority of the ATP is secured during respiration by the process of oxidative phosphorylation (Chapter 7) as opposed to the much more restricted ATP formation obtained via anaerobic substrate-level phosphorylation (Chapter 8). Facultative organisms display interesting regulatory effects when they are switched from aerobic to anaerobic conditions and *vice versa*, often characterized by the operation of alternative metabolic pathways.

Energy transduction via substrate-level phosphorylation is a scalar, cytoplasmic process involving phosphorylated and non-phosphorylated intermediates, whereas photophosphorylation (Chapter 10) and oxidative phosphorylation are membrane-associated vectorial processes in which ATP synthesis is driven by sequential redox reactions (described in Chapter 7) which energize the membrane by generating a proton gradient across it. The concept of membrane energization (achieved either by electron transfer reactions or by ATP hydrolysis) with an associated protonmotive force is central to the unifying chemiosmotic hypothesis of Peter Mitchell, discussed in Chapter 5. This hypothesis, which also affords explanations for reversed electron transfer (Chapter 9), the transport of solutes across membranes (Chapter 6) and cell motility (Chapter 5), has played a significant role in advancing the study of bioenergetics.

Although there is a proposed alternative to chemiosmosis to explain membrane energy transduction, namely the localized proton hypothesis of R.J.P. Williams, the vast majority of the published work has been interpreted in terms of the chemiosmotic hypothesis and it is this latter mechanism that has been adopted throughout this book.

A small group of chemotrophic bacteria, the chemolithotrophs, is able to secure the ATP needed for growth from the oxidation of inorganic compounds, using either oxygen or nitrate as electron acceptor and, generally, to synthesize all cellular components from carbon dioxide alone (Chapter 9). These organisms, which require reducing equivalents for the reduction of carbon dioxide, display a form of energy transduction additional to oxidative phosphorylation via customary forward electron transfer, namely reversed electron transfer from higher redox potential donors to $NAD(P)^+$, a process driven by the gradient of protons generated by forward electron transfer from the same donors to oxygen or nitrate.

The ATP derived from these different energy-transducing mechanisms is used by microbial cells to permit uptake of solutes (including essential nutrients) from the environment (Chapter 6), their growth and maintenance of vital functions (Chapter 4), motility (Chapter 5) and, in the case of certain organisms, synthesis of energy-storage compounds under appropriate conditions (Chapter 11). When adverse circumstances prevail and they are deprived of exogenous energy sources, their survival is aided by securing the ATP required for maintenance via endogenous metabolic processes that are described in Chapter 12.

While a discussion of microbial structure and composition lies outside the scope of this book (for details the reader is directed to specialist texts, e.g. Rogers *et al.*, 1980; Rogers, 1983), it is necessary nonetheless in this introduction to consider briefly those aspects of structure germane to an understanding of microbial energy conservation: the following résumé is therefore designed to fulfil this purpose.

1.2 Prokaryotic cell walls and membranes

The classical Gram stain, which segregates most bacteria into positively- or negatively-reacting groups, does so on the basis of differences in the composition of their outer layers. Gram-positive bacteria are characterized by a thick (15–80 nm), rigid outer wall composed mainly of peptidoglycan (a mucopeptide composed of a β-1, 4-linked polymer of N-acetylglucosamine and N-acetylmuramic acid with peptide cross links) which confers on the cell its distinctive shape. Within this wall is the cytoplasm, bounded by a thin (about 8 nm wide), cytoplasmic (plasma) membrane which is the principal permeability barrier to the cell; it carries the systems necessary for the transport of solutes, as described in Chapter 6, and also those for energy transduction (Chapter 7).

The Gram-negative organism has a more complex cell envelope. The peptidoglycan layer is thin (about 2 nm), and covered by an outer membrane (8 nm) composed mainly of phospholipid, lipoprotein and lipopolysaccharide; together these structures confer rigidity and the characteristic morphology of the organism. Beneath these outer layers is the cytoplasmic membrane, separated from them by the periplasmic space, which is less than 5 nm wide and contains certain solute-binding proteins, various hydrolytic enzymes and, in a few cases, electron carriers.

The present concept of membrane structure both in prokaryotes and eukaryotes is the fluid mosaic model of Singer and Nicolson (1972). This envisages a bilayer of phospholipids with their hydrophobic fatty acid chains oriented inwards and their hydrophilic heads facing outwards. Proteins are embedded in this bilayer to varying depths, some completely spanning the membrane, as for example in the case of the F_0 channel of ATP synthetase (p. 64); transmembrane proteins in the outer membrane of Gram-negative bacteria form pores which are called *porins* (Nikaido and Vaari, 1985). Prokaryotic membranes differ from mammalian, plant and eukaryotic microbial membranes in lacking sterols. The major lipid of the Gram-positive bacterial cytoplasmic membrane is phosphatidylglycerol together with phosphatidylethanolamine, whereas in Gram-negative membranes phosphatidylethanolamine is the principal component with smaller amounts of phosphatidylglycerol. The degree of fluidity of these membranes depends upon their liquid crystalline state, which is largely a function of the melting points of the constituent fatty acids of their phospholipids; since unsaturated fatty acids have lower melting points than those of the corresponding saturated acids the proportion of unsaturated fatty acids in a membrane is a determining factor (Ellar, 1978). Accordingly, thermophilic bacteria, which flourish at high temperatures, have a high proportion of long-chain saturated fatty acids in their membrane phospholipids, whereas the converse applies to the psychrophilic organisms which grow at low temperatures. Further, the proportion of unsaturated fatty acids in the membrane of a mesophile is inversely related to the temperature at which it is grown.

When sections of bacteria are examined microscopically after fixing and staining, many (especially Gram-positive species) display invaginations of their cytoplasmic membranes giving rise to goblet-like structures termed *mesosomes.* It is still controversial as to whether mesosomes are simply artefacts of the preparative methods employed. Nonetheless, internal specialized membranes are typical of photosynthetic micro-organisms and are known as *chromatophores* or *thylakoids.* They are derived from invaginations of the cytoplasmic membrane and carry photosynthetic pigments, although these compounds are not exclusively confined to such internal membranes.

The outer layers of bacteria may be likened to sieves; generally they permit relatively large molecules to penetrate their structure and, in the case of Gram-negative bacteria, enter the periplasmic space. The enzyme lysozyme in the presence of EDTA is able to hydrolyse the β, 1–4 linkages of the peptidoglycan layer of Gram-negative bacteria but does not appreciably affect the outer membrane; consequently the resulting protoplast is still enveloped by an outer membrane and is referred to as a spheroplast (see Figure 1.1*b*).

1.3 Eukaryotic cell walls and membranes

The cell walls of eukaryotic microbes, the fungi, differ significantly from those of prokaryotes. Generally, the principal structural macromolecule is chitin, a β-1, 4-linked polymer of *N*-acetylglucosamine; it is thus an analogue of cellulose (a β-1,4-linked polymer of D-glucose), the structural element of plant cell walls. Other polysaccharides are also present according to the organism.

Yeasts possess polysaccharide walls composed of an inner fibrillar network of glucan (mostly as a β-1, 3-linked polymer of D-glucose) embedded in an amorphous matrix of mannoproteins which extends to the outer layers. Chitin is also present, often associated with bud scars. In moulds such as *Penicillium* species there are two types of wall, distinguished by possession of α-glucose and a β-linked polymer of galactofuranose respectively. Common to both types is a β-glucan–chitin complex.

As with bacteria, it is possible to prepare protoplasts from fungal cells by degrading their cell walls in a hypertonic medium. The mixture of enzymes required for this purpose includes glucanases, chitinases and mannanases, which hydrolyse the respective wall polymers, and the most popular agent is the gastric juice of the snail *Helix pomatia.*

The cytoplasmic membranes of fungi do not, of course, resemble those of prokaryotes because they play no role in respiration and energy transduction, functions which are fulfilled by the mitochondria of the eukaryotes.

1.4 The use of membrane preparations: vesicles

Studies of the crucial role of membranes in energy-transduction have been greatly aided by techniques which enable the experimenter to prepare from mitochondrial or bacterial cytoplasmic membranes structures known as *vesicles*, i.e. topologically sealed units composed of membrane fragments. Although this book does not embrace experimental techniques, a brief résumé of the preparation of vesicles is offered here because of the importance of results obtained with them.

A characteristic property of all biological membranes after their rupture is the ability to reseal themselves into closed vesicles instead of remaining as open fragments. Consequently, since membranes are anisotropic, i.e they display 'sidedness', with certain proteins and enzymes being accessible to substrates from one side of the membrane only, the resealed vesicles may possess either the same orientation as the parent membrane (termed right-side-out) or they may show opposite orientation and are said to be inside-out. The method of preparation of the vesicles has been observed to govern their orientation. Thus those prepared by relatively mild treatment, e.g. osmotic shock of bacterial protoplasts or spheroplasts, are generally large (about 1 μm diameter) and right-side-out, while those produced by the more vigorous techniques of ultrasound or shearing in a French pressure cell are smaller (about 0.1 μm diameter) and are inside-out. These two types of vesicle have proved invaluable for the study of energy transduction in membranes since the inside-out vesicles can be used for investigations of oxidative phosphorylation and photosynthetic phosphorylation while the right-side-out vesicles are particularly useful for solute transport studies. It must be stressed, however, that these methods often yield preparations of vesicles that are not homogeneous and therefore care must be exercised in the interpretation of experimental findings unless the orientation of the vesicles has been established, for example by examination of preparations which have been negatively stained to reveal the stalked spheres of the inner membrane.

1.4.1 Mitochondrial membranes

The inner and outer membranes of mitochondria are approximately the same thickness (about 7 nm) although their composition differs significantly. The inner membrane contains approximately 21% of the total mitochondrial protein whereas the outer membrane accounts for only about 4% but has a higher proportion of lipid. The major lipids of both membranes

(a) Mitochondria

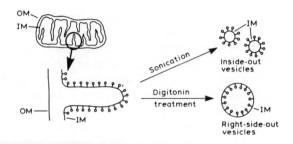

(b) Bacteria

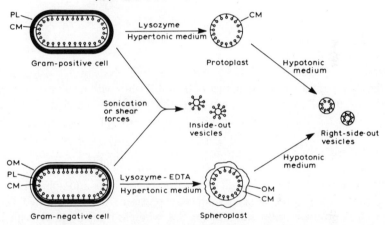

Figure 1.1 Diagrammatic representation of preparation of membrane vesicles: (*a*) from mitochondria by digitonin treatment (for right-side-out vesicles) or sonication (for inside-out vesicles); and (*b*) from bacteria, either directly (for inside-out vesicles) by sonication or shearing forces in a French pressure cell, or indirectly (for right-side-out vesicles) via protoplasts or spheroplasts by lysozyme or lysozyme-EDTA (for Gram-negative bacteria) treatment in hypertonic medium, followed by transfer to hypotonic medium.
Abbreviations: (*a*) OM, outer mitochondrial membrane; IM, inner mitochondrial membrane. (*b*) PL, peptidoglycan layer of the cell wall; CM, cytoplasmic membrane; OM, outer (lipopolysaccharide) membrane of Gram-negative bacterium; EDTA, ethylenediamine tetra-acetic acid.
Based on Prebble (1981) and Konings (1977).

are phosphatidylcholine and phosphatidylethanolamine. The matrix side of the inner membrane carries stalked spheres which are associated with the enzyme ATP synthetase (Chapter 5), and when mitochondria are treated as described below and shown in Figure 1.1*a*, it is principally the inner membrane which gives rise to the resulting vesicles.

1.4.2 Submitochondrial particles (vesicles)

Mitochondria of eukaryotic microbes are isolated by differential centrifugation after disruption of the cells by a relatively gentle method. The suitably washed mitochondrial preparation can then be treated in one of two ways (Figure 1.1a). Treatment with digitonin yields right-side-out vesicles which oxidize succinate and NADH but not other tricarboxylic acid cycle intermediates, and effect phosphorylation. Treatment with an ultrasonic oscillator produces inside-out (inverted) vesicles with the ATP synthetase and binding sites for respiratory chain substrates on the external surface. These two types of sub-mitochondrial particle have been used extensively in the investigation of electron transfer reactions and ATP formation.

1.4.3 Bacterial membrane vesicles

Treatment of bacteria with lysozyme causes hydrolysis of β-1, 4-linkages of the rigid peptidoglycan structure of their cell walls with resultant rupture and loss of cell contents. However, if the process is carried out in a hypertonic medium, i.e. of relatively high osmotic pressure, protoplasts or spheroplasts are obtained respectively from Gram-positive and Gram-negative organisms due to osmotic stabilization of the plasma membranes. If these bodies are then transferred to a hypotonic medium, the plasma membrane fragments into small pieces, the protoplasts and spheroplasts lose their integrity and the cell contents are released. As already noted, this procedure produces right-side-out vesicles, whereas if the original suspension of bacteria is sonicated or subjected to shearing forces in a French pressure cell, inside-out vesicles result (Figure 1.1b).

CHAPTER TWO

ADENINE NUCLEOTIDES AND THEIR
ROLE IN METABOLISM

2.1 The role of adenosine triphosphate

All living organisms function as open systems which require a continuous
supply of free energy for their growth and maintenance. This need may
be met either as chemical or physical (radiant) energy, in the case of
chemotrophs or phototrophs respectively. The cells of these organisms
transform the energy into the biologically useful form of adenosine
5'-triphosphate (ATP), which is the universal energy carrier of living matter,
mediating between energy-yielding and energy-utilizing reactions
(Figure 2.1).

ATP can hydrolyse in two different reactions which at pH7 have approxi-
mately the same standard free energy of hydrolysis ($\Delta G^{0'} \simeq -36\,\mathrm{kJ\,mol^{-1}}$).
In the first, the β, γ linkage is hydrolysed to yield adenosine 5'-diphosphate
(ADP) and inorganic phosphate (P_i), and in the second, hydrolysis of the

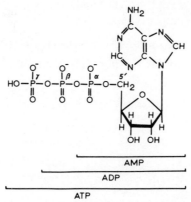

Figure 2.1   The structures of the three adenosine 5'-nucleotides.

9

α, β linkage occurs to furnish adenosine 5′-monophophate (AMP) and inorganic pyrophosphate (PP_i). To enable the cell to harness the free energy available from these hydrolyses the reactions must be coupled to other processes and must not occur as isolated events, otherwise the energy would be dissipated. Thus, as an example of the first type of reaction, we may consider the formation of glutamine from glutamate and ammonia, catalysed by *glutamine synthetase* according to the overall equation

$$\text{glutamate} + NH_4^+ + ATP \rightleftharpoons \text{glutamine} + ADP + P_i + H^+$$

For the purposes of thermodynamic calculation the reaction may be represented as two partial reactions, the first requiring an input of free energy ($\Delta G^{0'}$ positive) and being driven by the greater negative standard free energy of the hydrolysis of ATP:

$$\text{glutamate} + NH_4^+ \rightleftharpoons \text{glutamine} + H_2O + H^+ \qquad \Delta G^{0'} = +15.7\,\text{kJ}\,\text{mol}^{-1}$$
$$ATP + H_2O \rightleftharpoons ADP + P_i \qquad \Delta G^{0'} = -36.0\,\text{kJ}\,\text{mol}^{-1}$$

Net reaction:

$$\text{glutamate} + NH_4^+ + ATP \rightleftharpoons \text{glutamine} + ADP + P_i + H^+$$
$$\Delta G^{0'} = -20.3\,\text{kJ}\,\text{mol}^{-1}.$$

Although the mechanism of this reaction does in fact involve two stages, the first is actually the formation of enzyme-bound glutamyl 5-phosphate from glutamate and ATP and the second the release of the 5-phosphate associated with the formation of glutamine, namely

$$\text{glutamate} + ATP \rightleftharpoons \text{glutamyl 5-phosphate} + ADP$$
$$\text{glutamyl 5-phosphate} + NH_4^+ \rightleftharpoons \text{glutamine} + P_i + H^+$$

The phosphorylation of glucose to yield glucose 6-phosphate, coupled with the conversion of ATP to ADP and catalysed by *hexokinase*, affords a slightly different example. The reaction

$$\text{glucose} + ATP \rightleftharpoons \text{glucose 6-phosphate} + ADP$$

may be regarded (again for the purposes of thermodynamic calculation only) as two partial reactions:

$$\text{glucose} + P_i \rightleftharpoons \text{glucose 6-phosphate} + H_2O \quad \Delta G^{0'} = +12.0\,\text{kJ}\,\text{mol}^{-1}$$
$$ATP + H_2O \rightleftharpoons ADP + P_i \qquad \Delta G^{0'} = -36.0\,\text{kJ}\,\text{mol}^{-1}$$

Net reaction: $\text{glucose} + ATP \rightleftharpoons \text{glucose 6-phosphate} + ADP \quad \Delta G^{0'} = -24.0\,\text{kJ}\,\text{mol}^{-1}$

where the first reaction, requiring an input of free energy ($\Delta G^0 = +12.0\,\text{kJ}\,\text{mol}^{-1}$) is driven by the greater negative $\Delta G^{0'}$ of the second reaction ($-24.0\,\text{kJ}\,\text{mol}^{-1}$).

An example of the second type of ATP hydrolysis is found in the initial step of fatty acid biosynthesis where a fatty acid molecule is converted to its coenzyme A (CoASH) derivative at the expense of the pyrophosphate cleavage of ATP, thus:

$$R.COOH + CoASH + ATP \rightleftharpoons R.CO.SCoA + AMP + PP_i$$

The carboxyl group of the fatty acid forms a thioester linkage with the thiol group of coenzyme A. Another example is encountered in amino-acid activation for protein synthesis

$$R.CHNH_2.COOH + ATP \rightleftharpoons R.CHNH_2.COAMP + PP_i$$

in which the carboxyl group of the amino acid is bound in an anhydride linkage with the $5'$-phosphate group of AMP. It is generally found that the pyrophosphate cleavage of ATP is associated with the cellular processes where reversibility is not acceptable, for example in the biosynthesis of proteins, nucleic acids, glycogen and lipids, and to ensure that the reaction is effectively unidirectional *in vivo* the pyrophosphate formed is hydrolysed by an inorganic pyrophosphatase

$$PP_i + H_2O \rightarrow 2P_i$$

In recent years it has been recognized that in some micro-organisms pyrophosphate itself may serve as an energy mediator (see p. 156) and not simply dissipate its free energy of hydrolysis is an uncoupled process (Reeves, 1976).

ATP is an example of a so-called 'high-energy' compound: that is the standard free energy of hydrolysis of its anhydride bonds is significantly higher (i.e. more negative) than that of many organic phosphates (see Table 2.1) including AMP. Other high-energy compounds of importance in metabolism will be encountered in Chapters 3, 8 and 11. They include phosphoric-carboxylic anhydrides such as acetyl phosphate; N-phosphoguanidines, e.g. creatine phosphate and arginine phosphate; enol phosphates, e.g. phosphoenolpyruvate; and thiol esters, e.g. esters of coenzyme A, such as acetyl-SCoA and succinyl-SCoA.

The term 'high-energy phosphate bond' is frequently used in relation to such compounds and a 'squiggle' notation employed to denote that the hydrolysis of a given phosphate group is associated with a high standard free energy of hydrolysis. Thus ATP and ADP may be written as $A - P \sim P \sim P$ and $A - P \sim P$ respectively; it is a convenient formulation but one which must not be taken to imply that the energy resides in the bond and is released when the bond dissociates. Bond energies are always *positive*, i.e. energy must be utilized to break the bond, and consequently the concepts of a bond dissociation process yielding energy and

Table 2.1. Standard free energies of hydrolysis of some derivatives of phosphoric acid and coenzyme A.

Compound	$-\Delta G^{0'}$ (kJ mol^{-1})	pH	Temperature ($^\circ$C)
(a) *High-energy compounds*			
ATP^{4-} (β, γ linkage)	36.0	7.0	20
ATP^{4-} (α, β linkage)	36.9	7.0	20
ADP^{3-} (α, β linkage)	32.6	7.0	20
H$_2$P$_2$O$_7^{3-}$	34.7	7.0	20
Phosphoenolpyruvate^{3-}	55.6	7.0	20
Acetyl phosphate^{2-}	43.9	7.0	25
Creatine phosphate^{2-}	43.5	7.7	20
Arginine phosphate^{-}	49.4	7.7	20
Acetyl coenzyme A	34.3	7.0	–
Succinyl coenzyme A	32.6	7.0	–
(b) *Low-energy compounds*			
AMP^{2-}	9.6	7.0	20
Glycerol 1-phosphate^{2-}	9.2	8.5	38
Glucose 1-phosphate^{2-}	19.9	8.5	38
Glucose 6-phosphate^{2-}	12.6	8.5	38
Fructose 6-phosphate^{2-}	14.6	8.5	38

Note: $\Delta G^{0'}$ is the standard free energy change for the reaction at the specified pH with the compounds in solution at the standard state of unit activity (approximately 1M).

of energy being stored in the bond are physically meaningless. The standard free energy of hydrolysis is the relevant factor concerned and this is a property of the molecule as a whole (Dawes, 1980).

The factors which account for the relatively high standard free energies of hydrolysis of such compounds are essentially those which govern the relative instability of the reactant and the relative stability of the product. In this connection it is important to appreciate that, at pH 7, the convenient shorthand equations that we have used for the two different hydrolyses of ATP actually refer to reactions involving charged ionic species and the release of a proton, thus

$$ATP^{4-} + H_2O \rightleftharpoons ADP^{3-} + H_2PO_4^{2-} + H^+ \text{ and}$$
$$ATP^{4-} + H_2O \rightleftharpoons AMP^{2-} + H_2P_2O_7^{3-} + H^+$$

The reacting species (ATP^{4-}) is more negatively charged than the products, which have increased possibilities for resonance and hence greater stability than the reactant, separation of like charges occurs and a proton is released in the process. These factors contribute to the high negative standard free energy change on hydrolysis, although their relative contributions are still uncertain.

2.2 The phosphorylation potential and phosphate-group transfer potential

The free energy required to synthesize ATP from ADP and P_i has been termed the *phosphorylation potential*, ΔG_p, where

$$\Delta G_p = \Delta G^{0'} + 2.303 RT \log \frac{[ATP]}{[ADP][P_i]}$$

and the concentrations of ATP, ADP and P_i are those obtaining in the microbial cell (which are, of course, very much lower than the standard state concentration of 1M to which $\Delta G^{0'}$ refers). The reader should note that some authors define ΔG_p with reference to ATP hydrolysis, when its value will, of course, be negative.

It will be evident from the values of $\Delta G^{0'}$ of hydrolysis recorded in Table 2.1 that while the concept of high- and low-energy compounds has been a fruitful one for the understanding of bichemical energetics, there is not in fact a sharp line of demarcation between these groups. Lehninger considers the standard free energy of hydrolysis of ATP as constituting the mid-point of a thermodynamic scale of phosphorylated compounds. If this scale is arranged in descending order of $\Delta G^{0'}$ then those compounds highest in the scale undergo more complete hydrolysis at equilibrium than do those low in the scale, i.e. any compound higher in the scale than ATP would tend to donate its phosphate group to a phosphate acceptor molecule lower in the scale, for example ADP, provided an appropriate enzyme is available to catalyse the transfer. Lehninger introduced the term *phosphate-group transfer potential* as an arbitrary means of expressing the thermodynamic tendency or potential of the phosphate group of different phosphate compounds to undergo transfer; it is defined as the numerical value of $-\Delta G^{0'}$ when that parameter is expressed in $kJ \, mol^{-1}$.

The unique role of ATP in metabolism is explicable in terms of its bridging role between phosphate compounds with a high phosphate group potential and those having a low potential, so that it may function in transferring phosphate groups from the former to the latter compounds. Accordingly, ADP accepts phosphate groups from phosphate compounds of high potential in specific enzymic reactions and the ATP thereby formed then donates its terminal phosphate group enzymically to appropriate acceptor molecules, e.g. glucose, so raising their energy content; these are the low-potential compounds, e.g. glucose 6-phosphate. Consequently, position in the scale indicates the direction of enzymic transfer of phosphate under standard conditions.

At neutral pH, the adenine nucleotides or adenylates exist in solution as equilibrium mixtures of several polyanionic species, each of which can form complexes with divalent cations such as magnesium, Mg^{2+}, as well as with protons, e.g. $MgATP^{2-}$, $MgHATP^{-1}$, $MgADP^{-1}$. (It may be noted here that ATP^{4-} has a greater affinity for Mg^{2+} than either ADP^{3-} or AMP^{2-}.) Consequently knowledge of the equilibrium constants for the dissociation of these complexes is needed, and pH and ionic strength influence the concentrations of the species of adenylate present in solution and hence the free energy of hydrolysis of ATP (Blair, 1970). When these factors are taken into account it is found that magnesium ions cause a slight decrease in the standard free energy of hydrolysis of ATP to ADP and P_i. The intracellular concentration of available Mg^{2+} (i.e. not bound to cell components) in micro-organisms has been measured for relatively few species but for known examples is generally in the range of 1 to 10 mM; it thus normally exceeds the total adenine nucleotide concentration of most microbial cells which is in the range of 0.5 to 5.0 mM. (In contrast, we may note that in several types of mammalian cell the total adenylate pool is greater than the concentration of available Mg^{2+}.) In the presence and absence of 1 mM Mg^{2+} at pH 7, with ionic strength of 0.25 and at 25 °C, the standard free energy of hydrolysis ($\Delta G^{0\prime}$) for ATP to ADP and P_i is -31.8 and -35.6 kJ mol^{-1} respectively.

From the foregoing considerations it will be appreciated that the form in which ATP participates in reactions in the microbial cell is principally as the $MgATP^{2-}$ complex.

The adenylate system of the cell may therefore be said to comprise ATP, ADP, AMP, PP_i, P_i and Mg^{2+} ions, the adenylates being interconverted by the magnesium-dependent adenylate kinase, an enzyme of very high activity which catalyses the reversible reaction

$$ATP + AMP \rightleftharpoons 2ADP$$

or, more accurately

$$MgATP^{2-} + AMP^{2-} \rightleftharpoons MgADP^{-} + ADP^{3-}$$

with a $\Delta G^{0\prime}$ of approximately zero. Thus the enzyme may function to phosphorylate AMP to ADP or, in the reverse direction, to help maintain the ATP concentration in the cell by phosphorylating ADP at the expense of a second molecule of ADP, which is converted to AMP in the reaction (Bomsel and Pradet, 1968).

We have already noted that ATP (and hence also ADP and AMP) mediates between the catabolic and anabolic reactions of the cell. The former energy-yielding processes involve the conversion of suitable foodstuff molecules to smaller ones which can then serve as precursors for the biosynthesis of macromolecular components via the anabolic reactions, which require an energy input. ATP is produced in the catabolic sequences and utilized in the anabolic ones and it is for this reason that the adenylates are well fitted to regulate the metabolic economy of the cell. It has been discovered that many key regulatory enzymes in both anabolic and catabolic pathways are sensitive to ATP, ADP or AMP which serve as either positive or negative effectors for these allosteric proteins, i.e. either promoting or decreasing their activity, often by combining with the enzyme molecule at a location termed the *allosteric* or *regulatory* site, which is quite distinct from the catalytic or active site where substrate combination occurs. Protein conformational changes resulting from combination with an effector molecule modify the catalytic activity of the enzyme. Examples of such regulatory action are considered in Chapter 3 where the general principle may be discerned that ADP or AMP activates regulatory enzymes involved in catabolic pathways whereas ATP inhibits them; consequently energy-yielding substrates are degraded at maximum rate only when regeneration of ATP is necessary. Conversely, ATP activates and ADP or AMP inhibits many regulatory enzymes of biosynthetic sequences which are ATP-dependent. Some enzymes respond to the relative concentrations of ATP:ADP or ATP:AMP rather than to the absolute concentrations of individual adenine nucleotides.

2.3 The adenylate energy charge

The foregoing considerations emphasize the interdependence of ATP, ADP and AMP concentrations within living cells such that a change in one of them necessarily means changes in the concentrations of the other two. It is important therefore to have some unifying relationship to express these interactions. Accordingly, Atkinson (1968, 1977) introduced the concept of the *adenylate energy charge*, which enables the energetic state of a system, namely the total metabolic energy stored in the adenylate system, to be expressed quantitatively on a linear scale.

The total energy stored in the adenylate system is clearly proportional to the average number of anhydride-bound phosphate groups per adenosine moiety; this number varies between zero for AMP and two for ATP.

Adenylate energy charge (EC) is defined as

$$EC = \frac{[ATP] + \frac{1}{2}[ADP]}{[ATP] + [ADP] + [AMP]}$$

i.e. half of the number of anhydride-bound phosphate groups per adenosine (the half was introduced because it was preferable to have a linear scale running from 0 to 1, rather than from 0 to 2). Thus when all the adenylate of the cell is present as AMP the energy charge is zero, and when all is present as ATP, representing the maximum energetic state, the value is 1.0. On the vital assumption that adenylate kinase is highly active (the enzyme is ubiquitous in living cells), catalysing the interconversion of the three adenine nucleotides so that their concentrations are near equilibrium values at all times, their relative concentrations will vary with the energy charge of the system as shown in Figure 2.2. Inspection of this diagram reveals that, for any given point on the energy-charge scale, a metabolic process that utilizes ATP (anabolic) will move the system to the left and the concentrations of the individual adenylates will increase or decrease as indicated by their respective curves. Conversely, a catabolic process that regenerates ATP will move the system to the right. The function of the regulatory interactions in energy metabolism is to maintain the processes involved in ATP production and utilization in balance, and the concept of the adenylate energy charge has proved extremely useful in these studies.

We have already stressed that magnesium complexes of the adenylates

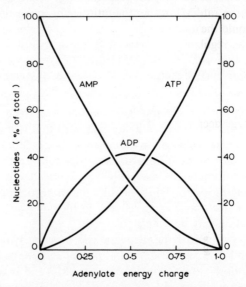

Figure 2.2 Relative concentrations of ATP, ADP and AMP as functions of the adenylate energy charge, assuming that the adenylate kinase reaction, 2ADP⇌ATP + AMP, was at equilibrium and the equilibrium constant to be 0.5.

are the actual reactive species involved in the adenylate kinase reaction and that ATP binds Mg^{2+} more firmly than does ADP or AMP. Consequently the apparent equilibrium constant for the kinase reaction depends on the Mg^{2+} concentration. While the general shapes of the curves in Figure 2.2 will not be affected by small changes in Mg^{2+} concentration, a change in the value of the apparent equilibrium constant will alter the value for the maximum concentration of ADP. For these reasons it will be apparent that Figure 2.2 gives a general rather than a precise representation of equilibrium conditions for adenylate kinase *in vivo*.

Atkinson has suggested that it is the adenylate energy charge which regulates the metabolic sequences producing and utilizing high-energy compounds. The relationship between the catabolic pathways regenerating (R) ATP and the anabolic ones utilizing (U) ATP as a function of energy charge is illustrated in Figure 2.3. The R-curve thus represents the response to energy charge of the catabolic reactions inhibited by ATP and/or activated by ADP or AMP, while the U-curve depicts the response of ATP-activated and ADP- or AMP-inhibited biosynthetic reactions.

It can be seen that enzyme activity is most responsive to energy charges of between 0.6 and 1.0 and that the curves intersect at an energy charge of approximately 0.85; at the point of intersection there is a metabolic steady state in which the rate of ATP regeneration is equal to the rate of ATP utilization. Should the energy charge vary from the value characteristic of the steady state, regulatory processes will come into play to restore the balance. Thus if the energy charge increases above the steady-state value the ATP-regenerating systems will be decelerated due to inhibition of their regulatory enzymes in response to the altered concentrations of ATP, ADP and AMP; likewise, the ATP-utilizing systems will be accelerated. Should the energy charge decrease then the converse situation will apply. In this

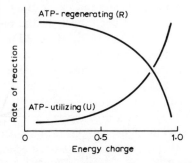

Figure 2.3. Theoretical responses to the energy charge of regulatory enzymes in metabolic sequences which lead to the regeneration (R) or utilization (U) of ATP. After Atkinson (1968).

way the adenylate system of the cell operates to ensure that the energy-yielding and utilizing sequences are maintained in a steady state for optimum economy, thereby preventing excessive supply or shortage of intermediary metabolites.

It must be appreciated, however, that the regulatory effect of adenine nucleotides on some enzymes is not necessarily due to their combination with allosteric sites. For example, enzymes that catalyse reactions involving the conversion of adenylates, such as kinases, will respond to the energy charge according to the relative affinity of their catalytic site for the substrate and the product adenine nucleotides.

Sensitivity of enzymes to feedback inhibition by products of a given metabolic sequence, as well as to energy charge, may modify the response shown in Figure 2.3. For both U- and R-type response curves the enzyme activity at a given energy charge will be lowered by an increasing concentration of a feedback modulator, thereby affecting the relative flux of metabolites through the particular sequences. Essentially this means that there is not one unique point of intersection of the two curves but rather a region of overlap. Thus overall metabolism can be regulated by small changes in energy charge and effector concentration and in an extremely flexible manner.

Eukaryotic microbes pose the problem of compartmentation of their adenylates between mitochondria and cytosol, thus complicating assessment of the effect of energy charge. Transport of ATP and ADP between cytosol and mitochondrial matrix occurs via a highly specific adenine nucleotide translocase which effects an exchange of intra-organelle and extra-organelle ATP and ADP on a 1:1 basis. AMP is not transported and the mitochondrial and cytosolic pools of AMP are therefore independent. Further, it has been found that the adenylate kinase of mitochondria resides in the intermembrane space and not in the matrix.

The respiratory activity of mitochondria in the presence of a suitable substrate is controlled by the availability of ADP and P_i within the mitochondrial matrix. Furthermore, in controlled mitochondria the ratio of ATP:ADP is lower than the extramitochondrial ratio, thus indicating that the mitochondrial energy charge is also lower. The influence of energy charge on the overall metabolic regulation of a eukaryotic cell is thus more difficult to interpret than with a prokaryote.

For the normal growth and metabolism of a microbial cell it is now generally recognized that the energy charge must be maintained within the range of 0.80 to 0.95. Within this range a small change in energy charge has relatively little effect on the ATP pool but a much greater effect on the AMP pool. For example, Knowles (1977) calculated that a decrease in energy charge from 0.92 to 0.80 caused only a 20% decrease in the ATP pool but a five-fold increase in the AMP concentration, resulting in a change in the ATP:AMP ratio from about 45:1 to about 8:1. A consequence of this amplification effect is that very small changes in energy charge can nonetheless have marked effects on the activity of enzymes that are regulated by adenine nucleotides.

One point is worthy of note, namely, that the energy charge is a unitless parameter and that, without additional information, it does not afford an indication of either the total adenylate pool size or the rate of turnover of ATP; both could vary widely without a change in energy charge. However, it is generally the case that enzymes which are regulated by energy charge are less sensitive to alterations in concentration of adenylates than to the ratio of ATP:ADP or ATP:AMP. Consequently micro-organisms are able to maintain their energy charge in response to environmental or metabolic stress by decreasing the adenylate pool size; this can be achieved either by excreting ATP or, particularly, AMP, or by the intracellular degradation of AMP to adenosine or adenine.

Measurements of the intracellular adenylate content of micro-organisms demand rapid sampling and quenching techniques because the ATP pool of growing organisms turns over many times per second and the action of adenylate kinase and ATPase can distort the profile of the individual adenylates. Further, because the total adenylate concentration in dilute suspensions of micro-organisms is low, very sensitive assay procedures are required. Usually either the luciferin–luciferase or the fluorimetric–enzymic method is used. In the latter case AMP and ADP have to be converted to ADP and ATP, respectively, before assay. Alternatively, thin-layer chromatography of ^{32}P-labelled nucleotides can be used. This technique is not quite so sensitive but it does have the advantage that other purine and pyrimidine nucleotides can be measured simultaneously.

The concept of the adenylate energy charge as a unifying hypothesis is not, however, without its critics. Knowles (1977), in a comprehensive review, points to the difficulty in assessing whether energy charge regulation determines metabolic homeostasis or whether its value is a consequence of such control. However, it is a parameter which has continued to be measured in studies of microbial energetics and we shall consider it subsequently in relation to metabolic regulation, growth and survival.

2.4 Fundamental mechanisms of ATP synthesis

There are two fundamental mechanisms by which ATP can be regenerated from ADP, namely substrate-level phosphorylation and electron transport (transfer) phosphorylation, the latter embracing both respiratory-chain and photosynthetic phosphorylation.

(1) Substrate-level phosphorylation. In this scalar, cytoplasmic process the formation of ATP from ADP is coupled to the enzymic transformation of a substrate containing a high-energy phosphoryl bond to its product. Thus, in the glycolytic sequence (p. 23), 1, 3-bisphosphoglycerate is converted by the enzyme phosphoglycerate kinase to 3-phosphoglycerate with simultaneous phosphorylation of ADP.

$$
\begin{array}{l}
O{=}C{-}O{-}PO_3^{2-} \\
\ \ \ | \\
H{-}C{-}OH \\
\ \ \ | \\
CH_2O.PO_3^{2-}
\end{array}
\ + \ ADP \rightarrow
\begin{array}{l}
O{=}C{-}O^- \\
\ \ \ | \\
H{-}C{-}OH \\
\ \ \ | \\
CH_2OPO_3^{2-}
\end{array}
\ + \ ATP
$$

A second example from glycolysis is the conversion of phosphoenolpyruvate to pyruvate catalysed by pyruvate kinase.

$$
\begin{array}{l}
CH_2 \\
\ \| \\
C{-}OPO_3^{2-} \\
\ | \\
CO_2^-
\end{array}
\ + \ ADP \ \rightarrow \
\begin{array}{l}
CH_3 \\
\ | \\
C{=}O \\
\ | \\
CO_2^-
\end{array}
\ + \ ATP
$$

A rather different kind of substrate-level phosphorylation reaction is encountered in association with the tricarboxylic acid cycle, an intermediate of which is the high-energy compound succinyl-SCoA. This molecule reacts with inorganic phosphate and guanosine diphosphate (GDP) under the influence of succinyl-SCoA synthetase:

$$\text{succinyl-SCoA} + P_i + GDP \rightleftharpoons \text{succinate} + GTP + \text{CoASH.}$$

(2) Electron-transport phosphorylation is a vectorial, membrane-associated mechanism in which the formation of ATP from ADP and P_i is coupled with the flow of electrons from a donor molecule to an acceptor molecule. This process commonly occurs in the respiratory chain and in photosynthetic reaction sequences (Chapters 7 and 10).

CHAPTER THREE

PATHWAYS OF GLUCOSE METABOLISM AND THEIR ATP YIELDS

Heterotrophic micro-organisms derive their energy from the catabolism of organic carbon substrates. Many bacteria are extremely versatile in their ability to use a variety of carbon sources, the pseudomonads for example growing on hydrocarbons, phenols, sugars and aliphatic amides. Most micro-organisms are rather more restricted in their biochemical potentialities although, with few exceptions, they are able to grow on glucose. It is convenient, therefore, to consider the metabolic pathways by which this compound can be metabolized to secure carbon skeletons and energy for biosynthesis.

The hexose molecule is degraded by a series of enzymic reactions to smaller molecules which can then be utilized for the synthesis of new cell material. Usually the molecule has to be phosphorylated before degradation, although some organisms are able to oxidize glucose before the phosphorylation step. In the course of these reactions ATP is generated for utilization in the biosynthetic reactions of the cell. Four major pathways for glucose metabolism in micro-organisms are currently known and these are usually related to the mode of life of the particular organism. The differences between these metabolic routes reside in the possession of key enzymes catalysing specific reaction sequences leading to the formation of triose phosphate and, indeed, all four pathways employ an identical series of enzymic reactions for converting triose phosphate to pyruvate. One may visualize, therefore, a central area of metabolism common to all pathways of glucose metabolism, with characteristic differences lying outside this core. As we shall see, the net ATP yield from glucose metabolism differs according to pathway and consequently it follows that these variations reside in the reactions which converge on the central zone, and on the further metabolism of pyruvate.

Micro-organisms, according to species, are able to live in the presence or absence of oxygen. Some are intolerant of one or other of these conditions and are classified as obligate aerobes or anaerobes, but others can tolerate both conditions and are referred to as facultative anaerobes. The metabolic pathway for glucose in a microbe must therefore be related to its mode of life. Under aerobic conditions oxygen will serve as the terminal electron acceptor and complete oxidation to carbon dioxide and water can occur, but under anaerobiosis products of glucose metabolism must serve as electron acceptors in a series of balanced oxido-reductions if glucose degradation and growth are to proceed. It is the further metabolism of pyruvate which assumes critical importance in these circumstances and the resulting metabolic end-products are characteristic of the given organism and of environmental conditions such as pH. These features will become apparent in the subsequent discussion of microbial glucose metabolism.

Thauer and his colleagues (1977) have pointed out that the energy metabolism of most organisms can be represented by a linear catabolic process with a constant ATP yield (Figure 3.1), citing homolactic fermentation and aerobic respiration as examples. The ATP yield of these processes is invariant. In contrast, the energy metabolism of many anaerobic microorganisms is branched (Figure 3.1), each branch leading to a different yield of ATP per molecule of glucose metabolized. Product formation by one branch would give a higher ATP yield than product formation by the other. The relative importance of a particular branch, and thus the ATP yield, will

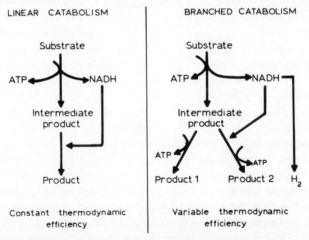

Figure 3.1. Comparison of a linear catabolic pathway having a constant thermodynamic efficiency with a branched catabolic pathway displaying a variable efficiency. After Thauer *et al.* (1977).

depend upon the conditions of growth; overall metabolism would be expected to be regulated to achieve the optimum growth yield under the given environmental conditions. Examples of such branched pathways with different ATP yields are considered in Chapter 8.

3.1 Pathways of glucose degradation

Four major routes of microbial glucose catabolism are currently recognized, namely (1) Embden–Meyerhof glycolysis, (2) pentose phosphate pathway, (3) Entner–Doudoroff pathway and (4) phosphoketolase pathway. The distinguishing features of these metabolic routes will now be discussed and their respective energy yields considered.

3.2 The Embden–Meyerhof glycolytic pathway

This metabolic sequence, which is characteristic of yeast fermentation, fungi and many bacteria, was the first to be discovered. It comprises some ten enzymes which catalyse the conversion of glucose (C_6) to pyruvate (C_3) under either aerobic or anaerobic conditions (Figure 3.2). Aerobically it normally operates in conjunction with the tricarboxylic acid cycle which effects the oxidation of pyruvate to carbon dioxide and water. Anaerobically, pyruvate, or products of its further (anaerobic) metabolism, must be reduced, for example with the formation of lactate or ethanol.

To initiate the glycolytic sequence the glucose molecule must be phosphorylated in the 6-position and the means by which this is accomplished depends on the type of organism. Yeast, fungi and some aerobic bacteria, including many pseudomonads, employ the enzyme *hexokinase*, which requires Mg^{2+} for activity and utilizes one molecule of ATP:

$$\text{glucose} + \text{ATP} \rightarrow \text{glucose 6-phosphate} + \text{ADP}$$

The reaction is essentially irreversible *in vivo*.

Facultative organisms, including *Escherichia coli* and streptococci which employ the phosphoenolpyruvate phosphotransferase system (p. 75) to transport glucose into the cell, effect the phosphorylation of glucose and other sugars during the entry process.

Glucose 6-phosphate is then isomerized by the enzyme *phosphohexose isomerase* to fructose 6-phosphate which becomes the substrate for the second phosphorylation step, catalysed by *phosphofructokinase*, a key enzyme of the pathway and the presence of which in an organism is taken as

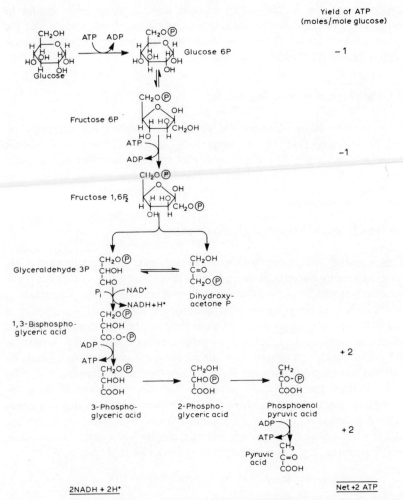

Figure 3.2. The Embden–Meyerhof glycolytic pathway. The overall reaction is given by the equation:

$$C_6H_{12}O_6 + 2NAD^+ + 2ADP + 2P_i \rightarrow 2CH_3CO.COOH + 2NADH + 2H^+ + 2ATP + 2H_2O$$

Anaerobically NADH is re-oxidized in the reduction of an organic fermentation product, e.g. pyruvate or acetaldehyde; aerobically via the electron transfer chain, yielding additional ATP in the process of oxidative phosphorylation.

good evidence for the operation of glycolysis. It resembles hexokinase in requiring ATP and Mg^{2+} and is irreversible *in vivo*:

fructose 6-phosphate + ATP → fructose 1,6-bisphosphate + ADP

Two molecules of ATP have thus been utilized in the formation of fructose 1, 6-bisphosphate, which is now split into two triose phosphates, dihydroxyacetone phosphate and glyceraldehyde 3-phosphate, by the action of *fructose bisphosphate aldolase*. This enzyme is important not only in the degradation of glucose but also in the reverse direction for the synthesis of hexose from non-carbohydrate precursors in the process of gluconeogenesis. Dihydroxyacetone phosphate is also an important linkage point for carbohydrate and lipid metabolism since it can be reduced to glycerol phosphate and used in lipid synthesis.

Although the triose phosphates are produced in equimolar amounts, it is glyceraldehyde 3-phosphate which undergoes further metabolism in glycolysis and dihydroxyacetone phosphate is converted to it under the influence of *triose phosphate isomerase*. Thus, in the succeeding reactions of this metabolic sequence, it must be borne in mind that two molecules of glyceraldehyde 3-phosphate traverse the pathway for each hexose molecule degraded. Further, it must be remembered that the reactions leading from glyceraldehyde 3-phosphate to pyruvic acid are common to all the pathways of microbial glucose metabolism.

The next step in glycolysis represents the first and only oxidation in the reactions leading to pyruvic acid. It is catalysed by an NAD^+-dependent thiol-containing enzyme, *glyceraldehyde 3-phosphate dehydrogenase*, and requires inorganic phosphate, which is incorporated in the product, 1, 3-bisphosphoglyceric acid. The reaction is very important in energetic terms for it is the first of the glycolytic sequence in which a high-energy phosphate compound is generated:

glyceraldehyde 3-P + NAD^+ + $P_i \rightleftharpoons$ 1, 3-bisphosphoglyceric acid + NADH + H^+.

The aldehyde group in the C-1 position is converted into an acyl phosphate, i.e. a mixed anhydride of phosphoric acid and a carboxylic acid, which has a high phosphate-group transfer potential, in the course of oxidation from the aldehyde to the acyl level. The reaction occurs via transfer of a hydride ion H^- to NAD^+ and the formation of a thioester, following combination of the aldehyde with the thiol group of a cysteine residue at the active site of the enzyme. The thioester is a high-energy intermediate which reacts with inorganic phosphate to yield 1, 3-bisphosphoglycerate, and the process is an example of substrate-level phosphorylation. Although

B

the standard free energy change of this reaction is positive ($+ 6.3 \, kJ \, mol^{-1}$), because the reaction is coupled to the next step in glycolysis, namely the transfer of the high-energy phosphate group of 1, 3-bisphosphoglycerate to ADP with synthesis of ATP, an overall negative $\Delta G^{0'}$ of $- 12.5 \, kJ \, mol^{-1}$ is achieved and the reaction proceeds from left to right.

The NADH formed must now be re-oxidized for metabolism to proceed. Aerobically this occurs via the respiratory electron transport system (p. 81); anaerobically it is coupled with the reduction of an organic compound and frequently with one which occurs later in the glycolytic pathway, namely pyruvate. Thus pyruvate is reduced to lactate and NADH is oxidized.

The high-energy phosphate group in the C-1 position of 1, 3-bisphosphoglycerate is now transferred to ADP by the action of *phosphoglycerate kinase*, a Mg^{2+}-dependent enzyme, leaving 3-phosphoglycerate as the other product:

$$\text{1, 3-bisphosphoglycerate} + \text{ADP} \rightleftharpoons \text{3-phosphoglycerate} + \text{ATP}$$

This is the first stage in glycolysis at which ATP is regenerated from ADP and is an example of substrate-level phosphorylation.

3-Phosphoglycerate is then converted to 2-phosphoglycerate by the enzyme *phosphoglyceromutase* which requires 2, 3-bisphosphoglycerate as cofactor and functions via a phosphorylated enzyme intermediate. Water is now removed from 2-phosphoglycerate by the action of *enolase*, which requires a bivalent metal ion, e.g. Mg^{2+}, Mn^{2+} or Zn^{2+}, to yield phosphoenolpyruvate:

$$\text{2-phosphoglycerate} \rightleftharpoons \text{phosphoenolpyruvate} + H_2O$$

The effect of dehydration is to produce another high-energy phosphate bond, this time an enol-phosphate, which is transferred to ADP under the influence of *pyruvate kinase*. This is the second ATP-yielding step of glycolysis and again is an example of substrate-level phosphorylation.

$$\text{Net reaction: glucose} + 2\text{ADP} + 2P_i \rightarrow 2 \text{ pyruvate} + 2\text{ATP}$$

The fermentation involves the expenditure of 2ATP (for the phosphorylation of glucose and of fructose 6-phosphate) and the generation of 4ATP (both triose phosphates yield 2ATP, as described above), giving a net yield of 2ATP in anaerobic glycolysis. However, in the case of micro-organisms which accumulate glycogen (Chapter 11) the initial breakdown step by *phosphorylase* is a phosphorolysis involving P_i and yielding glucose 1-phosphate directly:

$$(\text{glucosyl})_n + P_i \rightarrow (\text{glucosyl})_{n-1} + \text{glucose 1-phosphate}$$

The glucose moiety is thereby phosphorylated in the C-1 position without any expenditure of ATP. After conversion to glucose 6-phosphate by the action of *phosphoglucomutase*, it can enter glycolysis. Consequently, in these circumstances the net yield will be 3ATP per glucose unit converted to pyruvate.

3.2.1 Regulation of glycolysis

There are two principal functions of glycolysis, namely the provision of energy (directly and via intermediates which are oxidized in the tricarboxylic acid cycle) and the production of carbon building blocks for the biosynthesis of cell components. The term *amphibolic* has been applied to such dual-function pathways to distinguish them from the strictly catabolic and anabolic sequences of metabolism. The dual function of glycolysis imposes upon it a form of regulation rather more complicated than that observed with a strictly catabolic pathway involving only energy-linked control. An additional control feature, apparently unique to amphibolic pathways, is *precursor activation* which is, in effect, the opposite of feedback control where the last metabolite of a pathway inhibits the first enzyme of the sequence. In precursor control the first metabolite of the sequence activates the last enzyme of that pathway. Glycolysis provides the following example. Fructose 1, 6-bisphosphate may be regarded as the first metabolite of a sequence of reactions leading to pyruvic acid, the final enzymic step being the conversion of phosphoenolpyruvate to pyruvate, catalysed by pyruvate kinase. As we shall subsequently see, fructose 1,6-bisphosphate activates pyruvate kinase thus functioning as a positive 'feedforward' modulator. The general features of the regulation of glycolysis will now be considered.

When the adenylate energy charge of the microbial cell is low there must be some mechanism whereby the rate of glycolysis can be increased and, conversely, when the energy charge is high, of decreasing the rate of ATP formation. This control is achieved by modulating the rates of two key reactions in glycolysis, the phosphorylation of fructose 6-phosphate to fructose 1, 6-bisphosphate catalysed by phosphofructokinase, and the conversion of phosphoenolpyruvate to pyruvate effected by pyruvate kinase. Noteworthy is the fact that both these reactions are essentially irreversible, for it is a general principle of metabolism that enzymes catalysing such reactions are potential sites of control.

Phosphofructokinase is the most important regulatory site in glycolysis. While the general principles of its control are common to living systems, details differ between mammalian and microbial enzymes. Mammalian phosphofructokinase is inhibited by high levels of ATP, which lowers the affinity of the enzyme for its substrate fructose 6-phosphate. The inhibitory effect of ATP is reversed by AMP and thus when the ATP/AMP ratio is decreased, i.e. when the energy charge is low, the activity of the enzyme increases and glycolysis is stimulated. Control of the biosynthetic function of glycolysis is exerted by citrate, the initial product of acetyl-SCoA and oxaloacetate condensation in the tricarboxylic acid cycle, which inhibits phosphofructokinase. A high intracellular concentration of citrate means that biosynthetic precursors are abundant and deceleration of their further production by glucose breakdown is required.

Yeast phosphofructokinase resembles the mammalian enzyme in being stimulated by AMP and inhibited by both ATP and citrate but those bacterial enzymes so far studied display significant differences in their control. Thus the enzyme from the anaerobe *Clostridium pasteurianum* is not inhibited by ATP but is stimulated by ADP, while that from the facultative anaerobe *Escherichia coli* is activated by ADP and GDP but not by AMP, and is inhibited by phosphoenolpyruvate although not by citrate.

The discernible pattern of control is that phosphofructokinase is most active when the cell requires both energy and carbon skeletons for biosynthesis. Its activity is moderate when either energy or biosynthetic intermediates are required, and it is minimally active when neither is needed.

Pyruvate kinase in yeast is activated by fructose 1, 6-bisphosphate while the enzyme of *Bacillus licheniformis* is activated by its substrates, ADP and phosphoenolpyruvate, and inhibited by ATP, the latter inhibition being overcome by AMP.

As previously noted, not all micro-organisms phosphorylate glucose by the action of hexokinase, but for those that do there may be some element of regulation of this enzyme, for example inhibition by its product glucose 6-phosphate. However, glucose 6-phosphate is a branch-point intermediate common to all the different pathways of glucose catabolism. Entry to glycolysis depends upon its conversion to fructose 6-phosphate catalysed by phosphohexose isomerase, while entry to the pentose phosphate cycle, the Entner–Doudoroff and phosphoketolase pathways requires its oxidation to 6-phosphogluconate by glucose 6-phosphate dehydrogenase. Independent regulation of these routes of metabolism must clearly occur, therefore, by control of pacemaker reactions subsequent to that catalysed by hexokinase and, as discussed, in glycolysis these are the phosphofructose-kinase- and pyruvate-kinase-catalysed steps.

3.3 Glucogenesis

When micro-organisms grow on C_3 or C_4 compounds such as pyruvate or succinate as the sole source of carbon it is essential that they synthesize glucose from these substrates, a process termed glucogenesis (or gluconeogenesis). This poses an interesting problem in bioenergetics, because three reactions of glycolysis are virtually irreversible on account of their high negative ΔG, namely those catalysed by hexokinase, phosphofructokinase and pyruvate kinase, and it is not possible for the cell simply to reverse the glycolytic sequence in order to synthesize glucose. Alternative reactions have evolved to circumvent these irreversible reactions.

In *Escherichia coli* two enzymes enable phosphoenolpyruvate (PEP) to be synthesized when the organism grows on C_4 compounds, e.g. succinate or malate. These are *phosphoenolpyruvate carboxykinase* and *malate enzyme* which catalyse the respective reactions

$$HOOC.CH_2.CO.COOH + ATP \rightarrow CH_2{=}C(OPO_3H_2).COOH + CO_2 + ADP$$
oxaloacetate   PEP

$$HOOC.CH_2.CHOH.COOH + NADP^+ \rightarrow CH_3CO.COOH + CO_2 + NADPH + H^+$$
malate   pyruvate

If the carboxykinase is lost by mutation then growth on C_4 compounds and acetate is abolished, yet such mutants can still grow on C_3 compounds (pyruvate or lactate) because an Mg^{2+}-dependent enzyme *phosphoenolpyruvate synthase* is induced during growth on C_3 substrates and permits the reaction

$$CH_3CO.COOH + ATP \rightarrow CH_2{=}C(OPO_3H_2).COOH + AMP + P_i$$

to occur. Thus phosphoenolpyruvate is synthesized from pyruvate at the expense of the two high-energy bonds of ATP, one of which is preserved in the phosphoenolpyruvate formed.

Growth on C_3 compounds also demands that the organism synthesize C_4 intermediates from C_3 and CO_2 in order that the tricarboxylic acid cycle can produce both energy and intermediates for biosynthesis. The key enzyme involved is *phosphoenolpyruvate carboxylase* which catalyses the reaction

$$CO_2 + CH_2{=}C(OPO_3H_2).COOH \rightarrow HOOC.CH_2.CO.COOH + H_3PO_4$$

and therefore fulfils a replenishing function by enabling net synthesis of C_4 compounds; it belongs to the category of *anaplerotic* enzymes (see p. 46).

The irreversibility of the phosphofructokinase reaction is surmounted by the enzyme *fructose bisphosphatase* which hydrolyses the phosphate group from the C-1 position:

$$\text{fructose 1, 6-bisphosphate} + H_2O \rightarrow \text{fructose 6-phosphate} + P_i$$

The dependence of glucogenesis on this reaction is emphasized by the fact that mutants of *Escherichia coli* devoid of the enzyme are unable to grow on substrates such as acetate, succinate or glycerol and have an absolute requirement for hexoses.

The employment of these reactions in conjunction with the reversible steps of glycolysis enables the microbial cell to synthesize glucose 6-phosphate from C_3 and C_4 precursors. The control of glucogenesis also resides in these irreversible reactions. Fructose bisphosphatase from all known sources is inhibited by AMP and the phosphoenolpyruvate carboxylase of *Escherichia coli* is inhibited by malate and aspartate, high intracellular concentrations of these latter compounds signalling that synthesis of C_4-dicarboxylic acids is not needed. The carboxylase is activated by acetyl-SCoA, an increase in whose concentration can indicate a shortage of oxaloacetate and C_4 compounds generally; fructose, 1, 6-bisphosphate also activates.

Fructose 1, 6-bisphosphate occupies an important branch point of glycolysis and glycogenesis. An increase in the concentration of AMP inhibits both fructose bisphosphatase and ADPglucose pyrophosphorylase (p. 147), both being enzymes involved in glycogen synthesis. If the concentration of fructose 1, 6-bisphosphate increases then glycolysis is stimulated by its positive modulation of pyruvate kinase and phosphoenolpyruvate carboxylase, and glycogen synthesis by its activation of ADPglucose pyrophosphorylase. If further stimulation of glycolysis is not needed then phosphoenolpyruvate inhibits phosphofructokinase while simultaneously stimulating glycogen formation via activation of ADPglucose pyrophosphorylase.

It will be apparent from the foregoing discussion that since the enzymes phosphofructokinase and fructose bisphosphatase carry out opposing reactions, their stringent control is vital to the economy of the cell. Otherwise *futile cycles* would be established and the fructose 1, 6-bisphosphate formed from fructose 6-phosphate and ATP immediately hydrolysed to fructose 6-phosphate and P_i with an effective net energy loss of one ATP per cycle.

3.4 The pentose phosphate pathway

The pentose phosphate pathway of glucose metabolism, which diverges from glycolysis at the stage of glucose 6-phosphate, fulfils two major roles in living organisms, namely (1) the formation of pentose phosphates for biosynthesis of nucleotides and nucleic acids and (2) the generation of energy in the form of reducing power (as NADPH). The reductive steps of many biosynthetic pathways, e.g. fatty acid synthesis, require NADPH as reductant and not NADH. Since NADH is generated in all oxidative reactions of glycolysis and the tricarboxylic acid cycle except that catalysed by isocitrate dehydrogenase, the pentose phosphate pathway is important in supplying the required NADPH.

In this pathway three metabolic sequences may be discerned, the first being a series of oxidations leading to the pentose phosphate ribulose 5-phosphate and CO_2, the second the isomerization and epimerization of ribulose 5-phosphate to ribose 5-phosphate and xylulose 5-phosphate respectively, and the third a series of anaerobic rearrangements of the carbon skeletons of these pentose phosphates to yield hexose phosphates and triose phosphate. The triose phosphate is then either converted to pyruvate by reactions common to glycolysis and oxidized via the tricarboxylic acid cycle, or is converted to hexose phosphate by the action of fructose bisphosphate aldolase and fructose bisphosphatase. The overall reactions are shown in Figure 3.3.

The first enzyme of the pathway is an NADP-specific *glucose 6-phosphate dehydrogenase* which oxidizes glucose 6-phosphate to 6-phosphogluconate via 6-phosphogluconolactone which is hydrolysed by a lactonase enzyme. *6-Phosphogluconate dehydrogenase* then yields ribulose 5-phosphate, CO_2 and another NADPH. The enzymes *ribulose 5-phosphate-3-epimerase* and *ribose 5-phosphate isomerase* respectively convert ribulose 5-phosphate into xylulose 5-phosphate and ribose 5-phosphate. Thus the sum of the initial oxidative attack on glucose 6-phosphate may be written as

$$3 \text{ glucose 6-phosphate} + 6\text{NADP}^+ + 3\text{H}_2\text{O} \rightarrow 3 \text{ pentose 5-phosphate} + 3\text{CO}_2 + 6\text{NADPH} + 6\text{H}^+$$

giving a yield of pentose phosphate and reducing power (NADPH) in the ratio of 1:2. Should the microbial cell have precisely this requirement for these compounds then clearly the foregoing reactions alone will suffice. However, two other circumstances are likely to be encountered, namely conditions when the cell needs more NADPH than is furnished in that

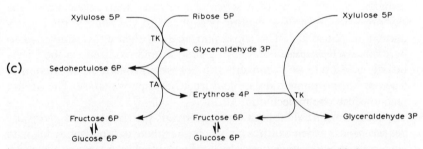

Figure 3.3. The pentose phosphate pathway of glucose metabolism. (*a*) Oxidative sequence of reactions. (*b*) Epimerization and isomerization reactions. (*c*) Non-oxidative sequence catalysed by transketolase (TK) and transaldolase (TA). Isomerization of the fructose 6P formed in sequence (*c*) to glucose 6P permits overall formulation of the pentose cycle as:

$$3 \, glucose \, 6P \rightarrow 2 \, glucose \, 6P + glyceraldehyde \, 3P + 3CO_2$$

ratio, and conversely more pentose phosphate relative to NADPH is required. The non-oxidative segment of the pathway, catalysed by *transketolase* and *transaldolase*, then comes into play.

Transketolase, a thiamine pyrophosphate-dependent enzyme, transfers a C_2-unit (effectively glycolaldehyde) from xylulose 5-phosphate to ribose 5-phosphate yielding sedoheptulose 7-phosphate (C_7) and glyceraldehyde 3-phosphate (C_3). Transaldolase, which has no known cofactor, then catalyses the transfer of a C_3-unit (effectively dihydroxyacetone) from sedoheptulose 7-phosphate to glyceraldehyde 3-phosphate to form fructose

6-phosphate (C_6) and erythrose 4-phosphate (C_4). The final stage involves transketolase again, acting upon xylulose 5-phosphate and erythrose 4-phosphate to produce glyceraldehyde 3-phosphate and a second molecule of fructose 6-phosphate. The net effect of these reactions is

2 xylulose 5-phosphate + ribose 5-phosphate $\rightleftharpoons$ 2 fructose 6-phosphate + glyceraldehyde
3-phosphate

Now if, in meeting the demands of the cell for NADPH, excess pentose phosphate is produced, it can be converted to fructose 6-phosphate, and so enter the glycolytic pathway. The glyceraldehyde 3-phosphate can yield pyruvate via reactions common to glycolysis and be oxidized in the tricarboxylic acid cycle. Some bacteria which can grow on glucose even though they lack key enzymes of the glycolytic and Entner–Doudoroff pathways, employ these reactions, e.g. *Brucella abortus* and *Thiobacillus novellus*. It is also possible under certain conditions that two glyceraldehyde 3-phosphate molecules can yield glucose 6-phosphate via the glucogenic reactions previously noted and then re-enter the pentose phosphate pathway. The overall stoichiometry in this case will be

6 glucose 6-phosphate + 12 $NADP^+$ + 7H_2O $\rightarrow$ 5 glucose 6-phosphate + 6CO_2
+ 12 NADPH + 12 H^+ + P_i

and, in effect, one molecule of glucose is completely oxidized for each turn of the cycle. This oxidative pentose cycle thus permits aerobic growth on hexose without the participation of the tricarboxylic acid cycle.

Conversely, if the demand for pentose phosphate exceeds that for NADPH, the need can be met by reversing these non-oxidative reactions and forming pentose phosphate from fructose 6-phosphate and glyceraldehyde 3-phosphate derived from glycolysis. The pentose phosphate pathway also furnishes erythrose 4-phosphate for the biosynthesis of aromatic amino acids.

In *Aspergillus* growing under nitrate-assimilating conditions the concentrations of pentose phosphate pathway enzymes increase relative to those found during growth on other nitrogen sources, an observation which correlates with the need for NADPH for nitrate reductase activity.

Assessments have been made of the proportion of glucose which is metabolized via the pentose pathway in various micro-organisms by the technique of radiorespirometry, extensively used by Wang and his colleagues. In *Escherichia coli* some 28% enters this pathway and glycolysis accounts for the remaining 72%. The non-oxidative segment of the pentose phosphate pathway involving transketolase and transaldolase is employed by many micro-organisms to degrade pentoses. After phosphorylation in the C-5 position, isomerization and epimerization, the pentose phosphates are converted to fructose 6-phosphate and glyceraldehyde 3-phosphate and gain entry to the glycolytic pathway.

3.4.1 *ATP yield in the pentose phosphate pathway*

The energy generated by the oxidative reactions of the pentose phosphate pathway effectively resides in the NADPH formed and will normally be utilized for reductive steps in biosynthesis. Since one mole of glucose 6-phosphate yields 12 NADPH via cycle oxidation, the theoretical yield per mole would be 36 ATP if it is assumed that (1) a transhydrogenase permits NADH formation, (2) the latter is reoxidized via the electron transport chain, and (3) the P/O ratio (p. 82) is 3.

3.4.2 *Regulation of the pentose phosphate pathway*

Control of glucose 6-phosphate entry to the pentose phosphate pathway is exerted on the first enzyme of the sequence, glucose 6-phosphate dehydrogenase. It is an allosteric enzyme commonly inhibited by ATP and by phosphoenolpyruvate in those organisms employing the Entner–Doudoroff pathway of glucose metabolism; hence it is controlled by the energy requirements of the cell. In *Azotobacter* ATP, NADPH and NADH inhibit the enzyme while the second enzyme of the sequence, an $NADP^+$-specific 6-phosphogluconate dehydrogenase, is also inhibited by NADPH and NADH, although unaffected by adenine nucleotides. There are both $NADP^+$-and NAD^+-specific glucose 6-phosphate dehydrogenases in *Pseudomonas fluorescens*; the former enzyme is not responsive to ATP and is associated with the pentose phosphate pathway whereas the NAD^+-linked enzyme, inhibited by ATP, is assigned to the major energy-yielding pathway, the Entner–Doudoroff sequence.

3.5 The Entner–Doudoroff pathway

Another important pathway of glucose metabolism first discovered by Entner and Doudoroff in *Pseudomonas saccharophila* was subsequently found to be widely distributed in bacteria. It is distinctive in that the metabolism of [1-^{14}C]glucose by this route yields carboxyl-labelled pyruvate instead of the methyl-labelled pyruvate characteristic of glycolysis. There are two key enzymes associated with the Entner–Doudoroff pathway, *6-phosphogluconate dehydratase* and *2-oxo-3-deoxy-6-phosphogluconate aldolase*, which operate after glucose 6-phosphate has first been oxidized to 6-phosphogluconate. The dehydratase removes the elements of water from 6-phosphogluconate to yield 2-oxo-3-deoxy-6-phosphogluconate, which is then split by the aldolase giving pyruvate and glyceraldehyde 3-phosphate; the latter compound is converted to a second molecule of pyruvate by reactions common to glycolysis (Figure 3.4).

This sequence of reactions occurs in aerobic bacteria such as pseudomonads and azotobacters and also in the anaerobic *Zymomonas*. It is important, too, when gluconate or other aldonic acids are utilized as substrates; gluconate kinase and the two key enzymes of the pathway are then induced. *Escherichia coli* degrades gluconate by the Entner–Doudoroff route and glucose via glycolysis.

3.5.1 ATP yield in the Entner–Doudoroff pathway

The anaerobic operation of this pathway is only half as efficient as anaerobic glycolysis, with a net yield of 1 mole of ATP per mole of glucose instead of 2, because only 1 mole of glyceraldehyde 3-phosphate is formed and oxidized. Although this gives 2ATP the net yield is only one because the glucose molecule must be phosphorylated.

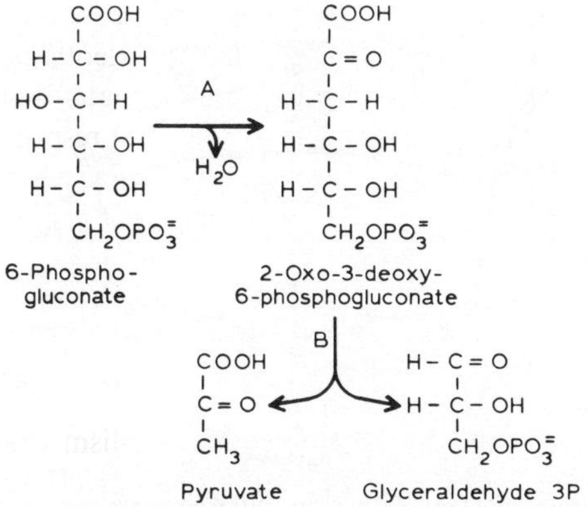

Figure 3.4.   The Entner–Doudoroff pathway. Enzyme A is 6-phosphogluconate dehydratase and enzyme B is 2-oxo-3-deoxy-6-phosphogluconate aldolase.

3.6 Non-phosphorylative oxidation

Many pseudomonads can also oxidize glucose to gluconate and 2-oxoglu-conate without prior phosphorylation, usually when excess of the substrate is available. A *glucose dehydrogenase* converts glucose to gluconate and *gluconate dehydrogenase* oxidizes gluconate to 2-oxogluconate. These two enzymes are located on the outside of the plasma membrane and thus the oxidations occur extracellularly in the periplasmic space. Gluconate and 2-oxogluconate are then taken up by specific transport systems, phos-phorylated by kinases and gain entry to the Entner–Doudoroff pathway; 2-oxogluconate-6-phosphate must first be reduced by an NADP-dependent reductase to 6-phosphogluconate (Figure 3.5). Generally the rate of oxidation of glucose exceeds the rate of uptake of the products and gluconate and 2-oxogluconate accumulate in the environment (Dawes *et al.*, 1976; Lessie and Phibbs, 1984).

The glucose dehydrogenases of the pseudomonads investigated, *Acinetobacter calcoaceticus*, and that of *Klebsiella aerogenes* induced under K^+-limited conditions, are quinoprotein enzymes possessing a pyrrolo-quinoline (PQQ) prosthetic group (Duine and Frank, 1981*a,b*). Of considerable significance is the discovery that the oxidation of glucose via these PQQ-dependent glucose dehydrogenases generates a protonmotive force and also supports secondary transport, both in whole cells and membrane vesicles prepared from them, indicating that the periplasmic non-phosphorylative oxidation of glucose yields energy to these organisms.

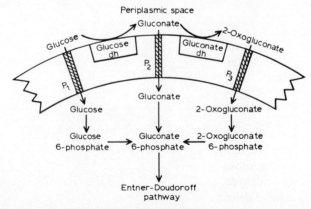

Figure 3.5. The extracellular (periplasmic) and intracellular pathways of glucose metabolism in *Pseudomonas aeruginosa*. P_1, P_2 and P_3 are the uptake systems for glucose, gluconate and 2-oxogluconate respectively. After Dawes *et al.* (1976).

3.7 The phosphoketolase pathway

The heterolactic bacteria are distinguished from the homolactic organisms by producing compounds other than lactate as major fermentation products of glucose. They also ferment pentoses with the formation of lactate and acetate. An example is *Leuconostoc mesenteroides* which ferments glucose according to the equation

$$glucose \rightarrow lactate + ethanol + CO_2$$

independently of glycolysis since the key enzymes of that pathway (phosphofructokinase and fructose bisphosphate aldolase) are absent (Figure 3.6). Radiochemical and enzymic evidence was obtained for a new pathway operating after glucose was phosphorylated to glucose 6-phosphate and converted to pentose 5-phosphate via the oxidative reactions of the pentose phosphate pathway (which involve NAD^+ in this organism). Xylulose 5-phosphate then becomes the substrate for a novel enzyme *phosphoketolase* which cleaves the molecule to acetyl phosphate and glyceraldehyde 3-phosphate. The enzyme requires thiamine pyrophosphate, inorganic phosphate, Mg^{2+} and a thiol compound for activity.

$$xylulose\ 5\text{-}phosphate + P_i \rightarrow acetyl\text{-}phosphate + glyceraldehyde\ 3\text{-}phosphate$$

Glyceraldehyde 3-phosphate is then converted to pyruvate by reactions common to glycolysis, yielding 2ATP, and the NADH formed in the oxidative step is reoxidized in the formation of lactate from pyruvate.

$$CH_3CO.COOH + NADH + H^+ \rightleftharpoons CH_3CHOH.COOH + NAD^+$$

The *Leuconostoc* enzyme is specific for xylulose 5-phosphate and does not attack fructose 6-phosphate although some organisms do have such an enzyme, e.g. *Acetobacter xylinum*, which forms erythrose 4-phosphate and acetyl phosphate.

When glucose is the fermentation substrate the acetyl phosphate is converted to acetyl-SCoA by the action of *phosphotransacetylase*

$$acetyl\text{-}phosphate + CoASH \rightleftharpoons acetyl\text{-}SCoA + P_i$$

and, to preserve the redox balance, the $2NADH + 2H^+$ generated in the formation of the pentose is reoxidized in the reduction of acetyl-SCoA to ethanol via acetaldehyde, catalysed by the enzymes *acetaldehyde dehydrogenase* and *ethanol dehydrogenase*:

$$CH_3CO.SCoA + NADH + H^+ \rightleftharpoons CH_3CHO + NAD^+ + CoASH$$
$$CH_3CHO + NADH + H^+ \rightleftharpoons C_2H_5OH + NAD^+.$$

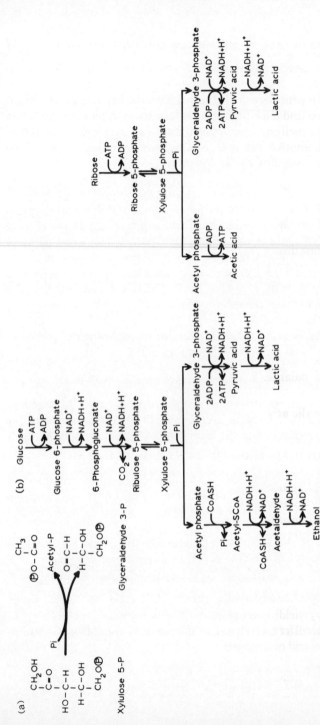

Figure 3.6. The phosphoketolase pathway. (*a*) The reaction catalysed by phosphoketolase, where P represents $-PO_3H_2$. (*b*) Conversion of glucose to lactic acid, ethanol and CO_2 and of ribose to lactic acid and acetic acid with respective net energy yields of 1ATP and 2ATP per molecule of substrate fermented, thus:

$$glucose + ADP + P_i \rightarrow lactic\ acid + ethanol + CO_2 + ATP + H_2O$$
$$ribose + 2ADP + 2P_i \rightarrow lactic\ acid + acetic\ acid + 2ATP + 2H_2O$$

The high-energy function of acetyl-SCoA is lost to the organism in the first reductive step.

However, when pentose is the fermentation substrate the initial oxidations do not occur and the redox balance is satisfied by lactate formation. The high-energy function of acetyl phosphate can then be conserved via acetate kinase in another example of substrate-level phosphorylation:

$$\text{acetyl phosphate} + \text{ADP} \rightleftharpoons \text{acetate} + \text{ATP}$$

Lipogenic yeasts (p. 78) have been found to possess phosphoketolase.

3.7.1 ATP yield via the phosphoketolase pathway

For the reasons just discussed, the net ATP yield via this pathway depends on the fermentation substrate. Glucose or pentose must first be phosphorylated at the expense of 1ATP and the reactions from glyceraldehyde 3-phosphate to pyruvate yield 2ATP. With pentose as the substrate a third ATP is derived from acetyl phosphate, making a net yield of 2 moles of ATP per mole of pentose, i.e. equivalent to glycolysis. However, the net yield is only 1ATP per mole of hexose as a consequence of the loss of the high-energy bond of acetyl phosphate by its reduction to ethanol in preserving the redox balance.

3.8 The tricarboxylic acid cycle

The various pathways of glucose metabolism previously discussed lead to the formation of pyruvic acid. Under aerobic conditions pyruvate is then oxidized via a cyclic process which is termed the *tricarboxylic acid cycle* or *citric acid cycle* (Figure 3.7). To gain entry to the cycle, pyruvate is first oxidatively decarboxylated to a C_2-unit in a multi-enzyme reaction catalysed by the *pyruvate oxidase* complex, which comprises three enzymes, five different cofactors and requires Mg^{2+} ions. The cofactors are thiamine pyrophosphate, lipoic acid, coenzyme A (CoASH), flavin adenine dinucleotide (FAD) and NAD. The net overall reaction may be written as

$$CH_3CO.COOH + CoASH + NAD^+ \rightarrow \text{acetyl-SCoA} + CO_2 + NADH + H^+$$

and it is an energy-yielding reaction by virtue of the reoxidation of NADH via the electron transfer chain being coupled to ATP synthesis (Chapter 7). The tricarboxylic acid cycle effects the oxidation of the two-carbon (acetyl) units to carbon dioxide and water and constitutes the most important single mechanism for the generation of ATP in aerobic micro-organisms. It is also of importance for the production of carbon skeletons for the synthesis of amino acids, e.g. aspartic and glutamic acids, and for porphyrins. The relative importance of the energy yielding and biosynthetic roles of the cycle depends principally upon whether or not the organisms are growing, as subsequently discussed.

Acetyl-SCoA condenses with oxaloacetate to yield citric acid and CoASH in a reaction catalysed by *citrate synthase.*

$$\text{acetyl-SCoA} + \text{oxaloacetate} \rightarrow \text{citrate} + \text{CoASH} + H_2O$$

This tricarboxylic acid is then isomerized to D-isocitrate via *cis*-aconitate, catalysed by *aconitate hydratase (aconitase)* which thus effects two distinct dehydration reactions, one involving a hydroxyl attached to a tertiary carbon atom and the other a hydroxyl attached to a secondary carbon atom.

The next step is the oxidative decarboxylation of isocitrate to 2-oxoglutarate under the influence of *isocitrate dehydrogenase*, an enzyme for which both NAD and NADP requirements have been established, depending upon the source of the enzyme.

$$\text{isocitrate} + NADP^+ \rightleftharpoons \text{2-oxoglutarate} + CO_2 + NADPH + H^+$$

The overall reaction is reversible and therefore permits CO_2 fixation to occur under appropriate conditions; in the tricarboxylic acid cycle it is one of the energy-yielding reactions because NAD(P)H is formed.

In a reaction analogous to the oxidation of pyruvate, 2-oxoglutarate is then oxidatively decarboxylated to succinyl-SCoA in a multi-enzyme sequence of reactions catalysed by the 2-*oxoglutarate oxidase* complex, in which process NAD is reduced.

$$\text{2-oxoglutarate} + \text{CoASH} + NAD^+ \rightarrow \text{succinyl-SCoA} + CO_2 + NADH + H^+$$

Succinyl-SCoA is a high-energy compound and is cleaved by *succinyl-SCoA synthetase (succinate thiokinase)* to succinate and CoASH with the concomitant formation of ATP from ADP and P_i.

$$\text{succinyl-SCoA} + ADP + P_i \rightleftharpoons \text{succinate} + \text{CoASH} + ATP$$

In the mitochondria of eukaryotic micro-organisms the corresponding reaction involves GDP in place of ADP, and the GTP formed subsequently transfers its terminal phosphate to ADP in a reaction catalysed by *nucleoside diphosphokinase.*

$$GTP + ADP \rightleftharpoons GDP + ATP$$

Some bacteria, however, are able to utilize either ADP or GDP.

Succinate is next oxidized to fumarate by *succinate dehydrogenase*, a

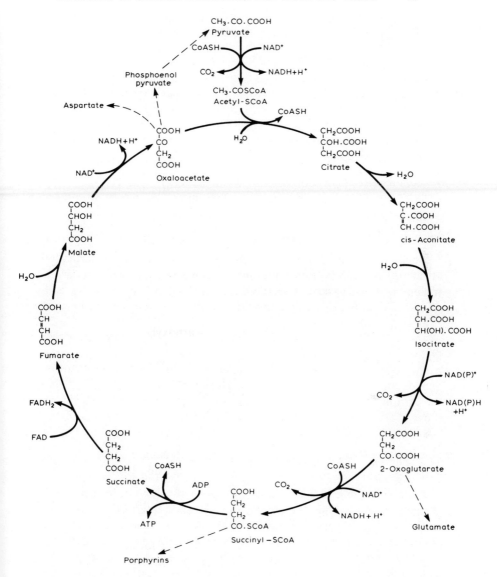

Figure 3.7. The tricarboxylic acid cycle. Biosynthetic routes from intermediates of the cycle are shown by broken lines.

flavoprotein (FP) enzyme which has a flavin adenine dinucleotide (FAD) prosthetic group and may be represented as E-FAD.

$$\text{succinate} + \text{E-FAD} \rightleftharpoons \text{fumarate} + \text{E-FADH}_2$$

The enzyme is associated with particles which also carry the necessary enzymes for the transfer of electrons to molecular oxygen via the cytochrome system and for this reason is often referred to as the *succinoxidase* system.

Fumarate is now reversibly hydrated to L-malate by the enzyme *fumarate hydratase* (*fumarase*)

$$\text{fumarate} + \text{H}_2\text{O} \rightleftharpoons \text{L-malate}$$

The final step of the cycle is the oxidation of L-malate to regenerate oxaloacetate by the action of an NAD-dependent *malate dehydrogenase*.

$$\text{malate} + \text{NAD}^+ \rightleftharpoons \text{oxaloacetate} + \text{NADH} + \text{H}^+$$

Although the equilibrium of this reaction greatly favours malate formation at neutral pH values, under the normal conditions of operation of the tricarboxylic acid cycle the oxaloacetate is removed by reaction with acetyl-SCoA, NADH is oxidized via the respiratory chain and therefore malate oxidation proceeds.

Thus a C_2-fragment produced in metabolism can be condensed with a C_4 acid, oxaloacetate, and by undergoing a series of reactions regenerates oxaloacetate and is itself effectively oxidized to two molecules of carbon dioxide and two molecules of water. However, it must be stressed that the carbon atoms of the regenerated oxaloacetate are not identical with those of the molecule of oxaloacetate that initiated the cycle.

3.8.1 The role of the tricarboxylic acid cycle

The tricarboxylic acid cycle fulfils two principal roles: (1) the generation of energy as ATP, and (2) the provision of carbon intermediates for biosynthesis. The relative importance of these functions depends upon whether or not the micro-organism is growing. Isotopic experiments have demonstrated that during exponential growth of *Escherichia coli* in glucose ammonium salts medium the major function of the tricarboxylic acid cycle is to furnish intermediates for synthesis, while glycolysis provides ATP. Cessation of growth causes a reversal of roles, the cycle now generating ATP and glucose being used for glycogenesis, leading to a deposition of glycogen in the cells (see Chapter 11).

Besides being important in the terminal oxidation of carbohydrates, the tricarboxylic acid cycle is also concerned in the oxidation of many other substances. Thus compounds that are degraded to acetyl-SCoA or to an intermediate of the cycle can be oxidized via this route. For example, fatty acids are degraded stepwise to acetyl-SCoA, and various amino acids are converted to the keto acids, pyruvic, oxaloacetic and 2-oxoglutaric acids, and thereby enter the cycle.

3.8.2 ATP yield from the tricarboxylic acid cycle

The generation of ATP by the tricarboxylic acid cycle occurs via the oxidative steps plus one substrate-level phosphorylation reaction. There are four oxidative steps within the cycle proper, namely the oxidation of isocitrate, 2-oxoglutarate, succinate and malate, and the substrate-level phosphorylation involving succinyl-SCoA. When pyruvate derived from glucose metabolism is the substrate for metabolism by this pathway then an additional oxidative reaction with concomitant energy yield is involved. Thus, if we consider the total energy yield in terms of molecules of ATP derived from the oxidation of pyruvate to carbon dioxide and water in the course of one turn of the tricarboxylic acid cycle in mitochondria, we can draw up a balance sheet. The ATP is generated during the transfer of electrons from NADH and $FADH_2$ to oxygen in the process of oxidative phosphorylation, as described in Chapter 7.

pyruvate + CoASH + NAD^+	$\rightarrow$ acetyl-SCoA + CO_2 + NADH + H^+	3ATP
isocitrate + $NADP^+$	$\rightarrow$ 2-oxoglutarate + CO_2 + NADPH + H^+	3ATP
2-oxoglutarate + CoASH + NAD^+	$\rightarrow$ succinyl-SCoA + CO_2 + NADH + H^+	3ATP
succinyl-SCoA	$\rightarrow$ succinate + CoASH	1ATP
succinate + FAD	$\rightarrow$ fumarate + $FADH_2$	2ATP
malate + NAD^+	$\rightarrow$ oxaloacetate + NADH + H^+	3ATP
	Total	15ATP

Hence in mitochondria 15 molecules of ATP are obtained by the oxidation of pyruvate in the cycle, 12 of these molecules arising from the cycle proper. The energy yield in prokaryotes is not always so high, however, because many of these organisms possess two, not three sites of energy conservation, as discussed in Chapter 7. Where only two energy coupling sites exist, the comparable yield from pyruvate in one turn of the cycle will be 10 molecules of ATP. Eukaryotic micro-organisms which employ the Embden–Meyerhof glycolytic pathway in combination with the tricarboxylic acid cycle would thus secure a total 38 molecules of ATP per molecule of glucose oxidized (30 ATP from the oxidation of two pyruvate molecules via the cycle, 2 ATP net by substrate-level phosphorylation in glycolysis and 6 ATP from oxidative phosphorylation of the 2 NADH generated in glycolysis) while many prokaryotes would obtain about 26 molecules of ATP. If these ATP yields are compared with those of anaerobic glycolysis (2 ATP) or metabolism of glucose via the Entner–Doudoroff (1 ATP) or phosphoketolase (1 ATP) pathways as previously discussed, the much greater efficiency of aerobic metabolism in terms of ATP yield is apparent and is reflected in molar growth yield studies, as described in Chapter 4.

3.8.3 Regulation of the tricarboxylic acid cycle

The tricarboxylic acid cycle is controlled principally at the point of entry of acetyl-SCoA, i.e. the first reaction of the cycle, catalysed by citrate synthase. Weitzman (1981) has found that the citrate synthases of many genera of Gram-negative bacteria are inhibited by NADH (which may be regarded as a 'product' of the operation of the cycle) whereas the enzyme from aerobic Gram-positive bacteria is not affected, although some of these enzymes are inhibited by ATP. Within the NADH-susceptible group, subgroups could be distinguished on the basis of the ability of AMP to reverse the inhibition by NADH; generally, reversal by AMP is characteristic of aerobes whereas non-reversal is typical of facultative organisms. Thus under conditions which lead to an accumulation of reducing power (NADH) and ATP, as for example when biosynthesis is curtailed, the overall operation of the tricarboxylic acid cycle will be inhibited in susceptible organisms, thereby conserving carbon and energy. Weitzman has demonstrated that the difference in regulatory behaviour is related to different enzyme structures. The NADH-sensitive citrate synthase is a large enzyme ($M = 250$k) composed of six subunits and subject to allosteric control, whereas the NADH-insensitive enzyme is smaller ($M = 100$k).

When facultative bacteria such as *Escherichia coli* grow under anaerobic conditions, synthesis of their 2-oxoglutarate dehydrogenase is repressed, and consequently the cycle becomes non-operative and is replaced by a fork, one limb leading reductively from oxaloacetate to succinate and the other oxidatively from citrate to 2-oxoglutarate (Figure 3.8). The enzymes of the oxidative limb, namely citrate synthase, aconitase and isocitrate dehydrogenase, are then required only for glutamate synthesis, and 2-

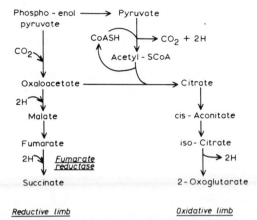

Figure 3.8.   The tricarboxylic acid cycle fork which operates in facultative bacteria under anaerobic growth conditions when 2-oxoglutarate dehydrogenase activity is absent. Succinate is formed by reduction of fumarate in a reaction catalysed by fumarate reductase, an enzyme induced only when anaerobic growth occurs.

oxoglutarate, the end product of this limb, effects control by inhibiting citrate synthase. The three enzymes are also present at lower levels under anaerobic than under aerobic conditions of growth. Furthermore, cyanobacteria lack 2-oxoglutarate dehydrogenase and in these organisms both 2-oxoglutarate and succinyl-SCoA inhibit citrate synthase, thereby effecting the control of carbon intermediates.

The catalysts of the reductive limb of the anaerobic fork are malate dehydrogenase, fumarate hydratase and fumarate reductase which replaces succinate dehydrogenase and is induced only under anaerobic conditions (p. 100). There are thus two reductive steps, requiring respectively NADH (malate dehydrogenase) and $FADH_2$ (fumarate reductase), which must be linked to suitable oxidative reactions in order to achieve the overall redox balance, and the coupled oxidation–reduction reaction must have a favourable equilibrium constant in order to promote succinate and 2-oxoglutarate formation.

Recently, Thauer (1982) has made the interesting discovery that certain sulphate-reducing bacteria (obligate anaerobes) oxidize acetate to CO_2 by means of a conventional tricarboxylic acid cycle. These organisms possess both 2-oxoglutarate dehydrogenase (although specific for ferredoxin as electron acceptor and not NAD^+) and succinate dehydrogenase, enzymes that, as we have observed, are normally absent under anaerobiosis. All the other enzymes of the cycle are present (except succinate thiokinase) although some have unusual properties: citrate synthase is AMP-dependent, fumarase is cold-labile and malate dehydrogenase couples with 1, 4-naphthoquinone rather than $NAD(P)^+$. For the cycle to operate, an anaplerotic mechanism (see the glyoxylate cycle, section 3.9) is needed to replenish those cycle intermediates that are continuously removed for biosynthesis. It is believed that the reductive carboxylation of acetyl-SCoA with reduced ferredoxin as electron donor catalysed by pyruvate synthase serves this purpose, with subsequent carboxylation of pyruvate to oxaloacetate.

Although the oxidation of pyruvate to acetyl-SCoA is not part of the tricarboxylic acid cycle, it does supply one of the substrates for citrate synthase and is itself also subject to regulation. Pyruvate dehydrogenase, the first enzyme of the pyruvate oxidase complex, is inhibited by acetyl-SCoA, the product of its action, by negative feedback inhibition. This enzyme is activated by AMP and inhibited when the adenylate energy charge is high.

3.9 The glyoxylate cycle

Since the tricarboxylic acid furnishes intermediates for the biosynthesis of amino acids, porphyrins etc. as well as generating ATP via oxidation of acetyl-SCoA, it follows that if the cycle is to continue functioning there must be some mechanism for replenishing the C_4 acids drained from it. This usually occurs by the carboxylation of pyruvate or phosphoenol-pyruvate to yield oxaloacetate

$$CH_3CO.COOH + CO_2 + ATP \rightarrow HOOC.CH_2.CO.COOH + ADP + P_i$$
$$CH_2{=}C(OPO_3H_2).COOH + CO_2 + H_2O \rightarrow HOOC.CH_2.CO.COOH + P_i$$

catalysed by the enzymes *pyruvate carboxylase* and *phosphoenolpyruvate (PEP) carboxylase* respectively. In *E. coli* the carboxylation of phosphoenolpyruvate seems to be the essential reaction, because only mutants lacking PEP-carboxylase fail to grow on pyruvate or its precursors unless the medium is supplemented with tricarboxylic acid cycle intermediates. However, this enzyme is absent from pseudomonads and *Arthrobacter* which possess pyruvate carboxylase instead. In both reactions it will be noted that the equivalent of one molecule of ATP is expended per molecule of oxaloacetate formed. These replenishment reactions are referred to as *anaplerotic*, from the Greek for 'filling up'.

Many micro-organisms are able to grow on acetate as the sole source of carbon and energy. Now if the tricarboxylic acid cycle is to continue to fulfil its dual role of providing energy and intermediates for biosynthesis under these conditions, there must be some mechanism for replenishing intermediates of the cycle from C_2-precursors. An anaplerotic pathway which operates in these circumstances is the *glyoxylate cycle*, the net effect of which is to yield one molecule of succinate from two molecules of acetate. This is achieved by the induction of two additional enzymes, *isocitrate lyase* and *malate synthase*, which catalyse the respective reactions:

$$\text{isocitrate} \rightleftharpoons \text{succinate} + \text{glyoxylate}$$
$$\text{acetyl-SCoA} + \text{glyoxylate} + H_2O \rightarrow \text{malate} + \text{CoASH}$$

These reactions, in conjunction with citrate synthase, aconitase and malate dehydrogenase, then permit the overall sequence:

$$\text{acetyl-SCoA} + \text{oxaloacetate} + H_2O \rightarrow \text{citrate} + \text{CoASH}$$
$$\text{citrate} \rightleftharpoons \text{isocitrate}$$
$$\text{isocitrate} \rightleftharpoons \text{succinate} + \text{glyoxylate}$$
$$\text{acetyl-SCoA} + \text{glyoxylate} + H_2O \rightarrow \text{malate} + \text{CoASH}$$
$$\text{malate} + \tfrac{1}{2}O_2 \rightarrow \text{oxaloacetate} + H_2O$$

Net reaction: $2 \text{ acetyl-SCoA} + \tfrac{1}{2}O_2 + H_2O \rightarrow \text{succinate} + 2\text{CoASH}$

The tricarboxylic acid cycle can thus continue to function normally for the oxidation of acetate and the production of intermediates for biosynthesis, while the glyoxylate cycle replenishes the cycle with succinate. It will be noted (Figure 3.9) that the glyoxylate cycle effectively bypasses the two oxidative decarboxylation reactions of the tricarboxylic acid cycle which are catalysed by isocitrate dehydrogenase and 2-oxoglutarate dehydrogenase. Further, the activation of acetate to acetyl-SCoA by acetyl-SCoA synthetase involves the conversion of ATP to AMP and PP_i which, because the pyrophosphate is subsequently hydrolysed to $2P_i$, means the equivalent expenditure of 2 molecules of ATP per molecule of acetate entering metabolism. Alternatively, acetyl-SCoA may be formed via acetate kinase and phosphotransacetylase reactions (p. 94) at the expense of one ATP.

3.9.1 Regulation of the glyoxylate cycle

The glyoxylate cycle is regulated principally by inhibition of activity of isocitrate lyase. Fine control in *E. coli* is exerted by phosphoenolpyruvate, which is a powerful non-competitive feedback inhibitor of the enzyme. Consequently accumulation of phosphoenolpyruvate would inhibit the

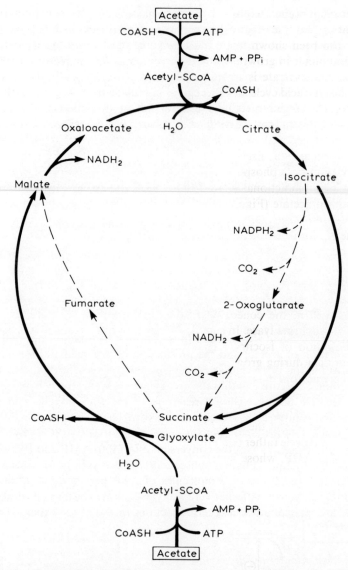

Figure 3.9. The glyoxylate cycle which permits net synthesis to occur when micro-organisms grow on C_2-substrates. The broken lines indicate reactions of the tricarboxylic acid cycle.

key enzyme of the anaplerotic sequence leading to phosphoenolpyruvate formation. In other micro-organisms succinate, glycollate and pyruvate have also been shown to inhibit isocitrate lyase. The presence of C_3 or C_4 compounds in growth media represses the synthesis of isocitrate lyase.

Because isocitrate is metabolized via isocitrate dehydrogenase in the tricarboxylic acid cycle and through isocitrate lyase in the glyoxylate cycle, the control of this metabolic branch-point is of great importance for micro-organisms growing on acetate. As we have seen, the glyoxylate cycle is essential under these conditions. It is now known that the isocitrate dehydrogenase of *Escherichia coli* is controlled by reversible phosphorylation, both phosphorylation and dephosphorylation being catalysed by a single bifunctional *kinase-phosphatase* enzyme which is induced by growth on acetate (Figure 3.10). The *E. coli* isocitrate dehydrogenase is unusual in that its phosphorylation can cause total inactivation.

The K_m for isocitrate dehydrogenase is very low $(1-2\,\mu M)$ whereas that for isocitrate lyase is over a thousandfold higher (3mM), exceeding the likely intracellular concentration of isocitrate. Thus inhibition of isocitrate dehydrogenase by phosphorylation during growth on acetate is believed to render the enzyme rate-limiting in the tricarboxylic acid cycle, causing an increase in the concentration of isocitrate and promoting its flux through isocitrate lyase. In support of this proposal, a higher intracellular concentration of isocitrate has been demonstrated during growth on acetate than during growth on glucose or glycerol.

Other effectors of isocitrate dehydrogenase may be classified into three groups: (a) precursors for biosynthesis, e.g. phosphoenolpyruvate, pyruvate, 2-oxoglutarate and 3-phosphoglycerate, which activate the phosphatase and inhibit the kinase, thereby promoting flux through the tricarboxylic acid cycle rather than the glyoxyate cycle; (b) adenylate nucleotides: AMP and ADP, whose increase in concentration reflects a decrease in

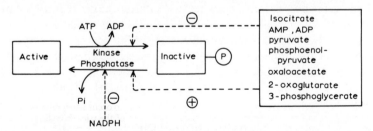

Figure 3.10. Regulation of isocitrate dehydrogenase of *Escherichia coli* during growth on acetate by reversible phosphorylation catalysed by a bifunctional kinase-phosphatase. The action of effectors is shown by broken lines. Based on Nimmo (1984).

the energy charge, activate the enzyme (ATP has an anomalous effect in activating the enzyme), as does anaerobiosis of acetate-grown *E. coli*, again probably mediated by the energy charge; (c) NADPH, which inhibits the phosphatase, and consequently isocitrate dehydrogenase, and may be regarded as an example of feedback inhibition of the tricarboxylic acid cycle by one of its end products (Nimmo, 1984). It must be emphasized, however, that there is no reason to believe that isocitrate dehydrogenase is a regulatory enzyme on carbon sources other than acetate or compounds that give rise solely to acetyl-SCoA.

3.10 Metabolism of other organic compounds

While the carbohydrate metabolic sequences already discussed, and the tricarboxylic acid cycle, constitute central pathways of metabolism, many micro-organisms, and especially pseudomonads, display remarkable versatility in the range of organic compounds they are able to degrade and utilize for growth. For example, soil organisms catabolize a wide array of materials derived from the plant kingdom, including the polymers lignin and cellulose and a variety of smaller molecules such as alkaloids, terpenes and flavonoids. Man-made pollutants of the environment like detergents can also be degraded. In all these cases the underlying metabolic strategy is the conversion of the substrate into metabolites which are members of one or other of the organism's central metabolic pathways. Such biochemical manipulation may involve short or relatively long reaction sequences, depending upon the structures of the compound to be degraded. These reactions may also necessitate an initial expenditure of energy and perhaps the induction of enzymes.

One particular attribute of aerobic micro-organisms, which greatly aids their ability to attack metabolically unreactive compounds such as hydro-carbons, is the possession of enzymes that catalyse the direct incorporation of atoms from molecular oxygen into the substrate molecules. Thus the terminal carbon of an alkane is oxidized by the introduction of a hydroxyl group under the influence of a *mono-oxygenase*, which requires reduced nicotinamide nucleotide as co-factor and therefore involves energy expenditure because of diversion of NADH from oxidative phosphorylation.

$$R{-}H + O_2 + NADH + H^+ \rightarrow R{-}OH + NAD^+ + H_2O$$

One atom of the oxygen molecule is incorporated into the alkane and the other into water. The resulting alcohol is then oxidized to the corresponding acid which, as an acyl-SCoA, can enter the fatty acid β-oxidation spiral. Saturated ring compounds are likewise inert until a hydroxyl group has been introduced; further oxidation may lead to ring fission and the formation of carbon fragments that can be degraded. The mono-oxygenase may be seen, therefore, to prepare the ring for fission.

Dioxygenases introduce both atoms of molecular oxygen into substrates containing a benzene nucleus, as in the *ortho* (1, 2) or *meta* (2, 3) cleavage of catechol (1, 2-dihydroxybenzene).

$$\text{(benzene ring)}\underset{\text{COOH}}{\overset{\text{COOH}}{}} \xleftarrow{O_2} \text{(benzene ring)}\underset{\text{OH}}{\overset{\text{OH}}{}} \xrightarrow{O_2} \text{(benzene ring)}\overset{\text{CHO}}{\underset{\text{OH}}{\text{COOH}}}$$

Subsequent metabolism yields acetyl-SCoA via the *ortho* route, and pyruvate and acetaldehyde via the *meta* pathway, and hence convergence, by two quite separate sequences, on the tricarboxylic acid cycle. For an excellent discussion of these and other metabolic pathways in aromatic metabolism leading to pyruvate and/or an intermediate of the tricarboxylic acid cycle, as well as consideration of carbohydrate and amino-acid catabolism, the reader is referred to Dagley (1978).

It should be noted that in those catabolic sequences initiated by expenditure of energy in the form of reduced nicotinamide nucleotides, and which may involve a fairly long sequence of reactions not linked to ATP generation before pyruvate or an intermediate of the tricarboxylic acid cycle is formed, an overall net ATP gain ultimately accrues to the organism via the terminal oxidation cycle.

CHAPTER FOUR

ENERGETICS OF MICROBIAL GROWTH

4.1 Growth yield coefficients

A micro-organism needs sources of carbon, energy, nitrogen, phosphorus, sulphur and a variety of mineral elements in order to grow. In the process of growth these various compounds and elements are converted into cell material at the expense of the energy source which, according to the major physiological groups, may be radiant energy or either inorganic or organic substrates. The ATP which is generated from these energy sources is required for (1) uptake of nutrients from the medium; (2) the biosynthesis from simpler substrates of the intermediates (monomer units) required for polymer synthesis, e.g. amino acids, nucleotides, fatty acids, sugar phosphates; (3) polymerization of the monomers to yield the polymers of the cell, e.g. macromolecules such as proteins, nucleic acids, polysaccharides and lipids; and (4) purposes independent of growth, such as the regulation of internal pH and osmotic pressure, resynthesis of degraded cell constituents (turnover) and motility, usually referred to as the maintenance energy requirement.

Monod (1942) first demonstrated a linear relationship between the total amount of growth (G) of a micro-organism and the quantity of energy and carbon source in the medium (C), i.e.

$$G = KC$$

where K is a constant, the *yield coefficient*. Thus if G is expressed in grams per litre and C as a molar concentration ($mol\,l^{-1}$) then K will be the g dry weight of organism supported by 1 mole of the substrate: this value is known as the *molar growth yield* and is recorded as Y with a subscript to indicate the carbon and/or energy source, e.g. $Y_{glucose}$ (often abbreviated to Y_{glc}).

Bauchop and Elsden in 1960 reported experiments with bacteria and yeast grown anaerobically, i.e. under conditions where the ATP yield by substrate-level phosphorylation could be accurately charted, and they concluded that the amount of growth was directly proportional to the ATP that could be derived from the catabolism of the carbon source. Thus organisms employing anaerobic Embden–Meyerhof glycolysis, and securing 2ATP per mole glucose degraded, displayed approximately double the molar growth yield of bacteria using the Entner–Doudoroff pathway from which only 1 ATP per mole of glucose can be derived (see pp. 24 and 35). The concept of Y_{ATP} was introduced, defined as the g dry weight of organism produced per g mol of ATP; their data showed a remarkable constancy for Y_{ATP} of about 10.5, leading to the proposal that Y_{ATP} was a biological constant. However, subsequent work with a much wider variety of organisms has shown that Y_{ATP} is not a constant, and values varying from 4.7 for *Zymomonas mobilis* to 20.9 for *Lactobacillus casei* have been reported. Clearly such calculations depend upon a precise knowledge of the pathway of catabolism of a substrate and its ATP yield. Where, for instance, there are two alternative pathways for the further catabolism of fermentation intermediates and these pathways are associated with different ATP yields, then the proportion entering each sequence will obviously affect the overall ATP yield. An example is provided by *Streptococcus faecalis* which displays Y_{glc} values that are dependent upon the glucose concentration; high glucose concentrations give lower Y_{glc} values than do low glucose concentrations. The explanation lies in the fact that at high glucose concentrations the organism carries out a homolactic fermentation, i.e. yielding solely lactate, with 2ATP generated per mole of glucose, whereas at low glucose concentrations products other than lactate, e.g. acetate and ethanol, are also formed and these reactions generate additional ATP. Further, under anaerobic conditions the presence in the medium of alternative electron acceptors such as nitrate, sulphate or fumarate can exert a profound influence on the ATP yield and there must be an awareness of these factors.

Although the earlier work on growth yields was carried out with batch cultures, continuous cultivation is now generally employed to permit growth to be secured in a controlled environment and at a rate determined by one specific limiting nutrient. Such studies have demonstrated that Y_{glc} and Y_{ATP} are dependent upon the specific growth rate (μ)[1] and only when μ approaches its maximum value (μ_{max}) are Y_{ATP} values as high as those

---

[1] $\mu = \dfrac{1}{x} \cdot \dfrac{dx}{dt}$ where x is the mass of micro-organism.

for the corresponding batch cultures obtained. The differences are most marked when the energy source is not the limiting factor and the discrepancy is greatest at very low growth rates because the proportion of the total energy production devoted to purposes other than the formation of new cell material is then at its highest.

4.2 Theoretical maximum growth yields: Y_{ATP}^{max}

It is possible to carry out theoretical calculations of the ATP required for the synthesis of cell constituents if the macromolecular composition of a given organism is known (Stouthamer, 1979). Because environmental factors such as the composition of the growth medium, availability of oxygen and the rate and temperature of growth affect the composition of microbial cells, it is important that the analytical data employed in the calculation are relevant to the cultural conditions although, as we shall see, variations in macromolecular composition, other than of storage compounds such as polysaccharide and poly-β-hydroxybutyrate, have surprisingly little effect on the ATP requirement. Very high growth yields are manifest when microbial cells contain large amounts of storage polymers. Various workers have calculated the ATP required for the synthesis of each cell constituent from the components of defined inorganic salts media providing a carbon source and ammonium or amino acids as the nitrogen source, and have used the information to compute the theoretical maximum Y_{ATP} or Y_{ATP}^{max}. Inspection of the values obtained for *Escherichia coli* (Table 4.1) reveals that there is not a great difference when preformed monomers (amino acids) replace ammonia in a glucose inorganic salts medium, e.g. Y_{ATP}^{max} of 31.9 and 28.8 respectively, indicating that there is only a small energy requirement for amino acid synthesis from glucose and ammonium. The actual ATP requirement for the synthesis of amino acids from glucose and ammonium represents the difference between the value recorded for protein synthesis from amino acids (i.e. energy used solely for polymerization of monomers, column A) and that for protein synthesis from ammonium (column B), namely $(205.0 - 191.4 = 13.6) \times 10^{-4} \, mol \, g^{-1}$ cells. In a similar manner the ATP needed for the synthesis of other monomers can be obtained by subtracting the corresponding values in column A from those in B and, for growth on pyruvate, column C from column D (Table 4.1).

The importance of the carbon source for Y_{ATP}^{max} is amply illustrated by the data of Table 4.1. Autotrophic growth with CO_2 displays a Y_{ATP}^{max} of only 6.5, while heterotrophic growth with acetate has Y_{ATP}^{max} of 10.0 and

Table 4.1. ATP requirement for the formation of microbial cells from various carbon sources in the presence or absence of amino acids and nucleic acid bases (Stouthamer, 1977, 1979).

Macromolecule synthesized	Macromolecule content g/100 g cells	ATP requirement ($10^4 \times$ mol/g cells formed)						
		A	B	C	D	E	F	G
Polysaccharide	16.6	20.6	20.6	71.8	71.8	51.0	92.0	195.0
Protein	52.4	191.4	205.0	191.4	339.4	285.0	427.0	907.0
Lipid	9.4	1.4	1.4	27.0	27.0	25.0	50.0	172.0
RNA	15.7	24.0	43.7	46.2	71.2	70.0	101.0}	212.0
DNA	3.2	5.7	10.5	9.9	15.9	13.0	19.0}	
Turnover mRNA		13.9	13.9	13.9	13.9	13.9	13.9	13.9
Total		257.0	295.1	360.2	539.2	457.9	702.9	1499.9
Transport		57.4	52.0	115.5	200.0	200.0	306.0	52.0
Total ATP requirement		314.4	347.1	475.7	739.2	657.9	1008.9	1551.9
Gram cells per mol ATP		31.9	28.8	21.0	13.5	15.4	10.0	6.5

The cell composition is that reported for *Escherichia coli* by Morowitz.
Media: A, glucose, amino acids and nucleic acid bases; C, pyruvate, amino acids and nucleic acid bases; B and D–G, inorganic salts plus (B) glucose, (D) pyruvate, (E) malate, (F) acetate and (G) carbon dioxide.

malate 15.4, compared with the glucose value of 28.8. These differences reflect the greater energy expenditure for synthesis of monomers and for transport processes when growth occurs with simpler carbon compounds than glucose. (It should be noted, however, that the values recorded for transport are likely to be over-estimates because of uncertainties in our knowledge of some of the processes involved.) Likewise, nitrogen-fixing organisms exhibit significantly lower molar growth yields when growing on molecular nitrogen than on ammonium due to the high energy expenditure in the reduction of nitrogen, e.g. *Klebsiella pneumoniae* needs about 15 mol ATP per mol ammonia formed. The greater need for reducing equivalents when growth occurs on nitrogen also means that there is a lower yield of ATP via oxidative phosphorylation per mole of substrate metabolized.

The highest biosynthetic ATP requirements are for protein, RNA and DNA which, for growth in a glucose-inorganic salts medium, lie in the range of about 33 to 39 mmol ATP per g macromolecule synthesized, compared with values of about 12.4 for polysaccharide and only 1.5 for lipid. Consequently if an organism has a high content of polysaccharide or lipid (see Chapter 11) the theoretical Y_{ATP}^{max} will be much higher than for the corresponding organism without these storage materials.

4.3 Aerobic growth yields: Y_{O_2} and $Y_{O_2}^{max}$

Growth of facultative micro-organisms under aerobic conditions (or anaerobic growth in the presence of external electron acceptors such as nitrate or fumarate) is faster than under anaerobiosis, the ATP yield is higher and greater molar growth yields are recorded. For example, *Proteus mirabilis* displays an anaerobic Y_{glc} of 14.0, an aerobic value of 58.1, and 30.1 when grown anaerobically with nitrate. However, aerobic growth presents problems of interpretation of yield data because in many cases the ATP yield in oxidative phosphorylation is not accurately known (p. 85). The dilemma presented is that one needs knowledge of either the $P/2e^-$ quotient to determine Y_{ATP}^{max} or, conversely, Y_{ATP} or Y_{ATP}^{max} must be known to determine the $P/2e^-$ quotient.

Measurement of the molar growth yield per mole of oxygen utilized during growth (Y_{O_2}) at a series of different growth rates enables $Y_{O_2}^{max}$, i.e. the value corrected for maintenance respiration, to be calculated from the equation

$$\frac{1}{Y_{O_2}} = \frac{1}{Y_{O_2}^{max}} + \frac{m_0}{\mu}$$

where m_0 is the maintenance respiration rate [mol O_2 mg^{-1} (dry weight)h^{-1}], using a plot of $1/Y_{O_2}$ versus $1/\mu$. Division of $Y_{O_2}^{max}$ by $2Y_{ATP}^{max}$ gives the $P/2e^-$ quotient while $Y_{O_2}^{max}/Y_{ATP}^{max}$ affords the $\rightarrow H^+/O$ quotient (see p. 89 for an explanation of this term), which can be determined experimentally with cell suspensions or extracts. However, in turn, this requires an assumption concerning the number of protons translocated per $2e^-$ transferred in a segment of the respiratory chain; while $2H^+$ per conservation site is commonly accepted, the evidence is not unequivocal. It must also be remembered that measurements of Y_{O_2} and $\rightarrow H^+/O$ do not embrace the significant amounts of ATP generated by substrate-level phosphorylation.

4.4 Discrepancies between experimental and theoretical growth yields

Large discrepancies were discovered between experimentally determined growth yields and theoretical values under both aerobic and anaerobic conditions. These differences might be attributed to (1) the energy requirements for maintenance of the cells, including the need to maintain the cytoplasmic membrane in an energized state in the face of energy dissipation due to some leakiness of the membrane towards protons, and to (2) a certain degree of uncoupling between growth and energy generation because the rate of ATP production in catabolism exceeds the rate of its consumption in anabolism. These factors will now be considered.

4.4.1 Energy of maintenance

Essential cellular processes, whether chemical or mechanical, e.g. the turnover of macromolecules, osmotic regulation, maintenance of intracellular pH and motility, require energy for their performance. This *energy of maintenance* is defined as the energy consumed for purposes other than the production of new cell material. It has usually been measured by examining the effect of specific growth rate ($\mu = 1/x.\mathrm{d}x/\mathrm{d}t$) on the molar growth yield in continuous culture and applying an equation introduced by Pirt in 1965:

$$\frac{1}{Y_{glc}} = \frac{m_s}{\mu} + \frac{1}{Y_{glc}^{max}}$$

where m_s is the maintenance coefficient [g glucose utilized g^{-1} (dry

C

weight)h^{-1}] and Y_{glc}^{max} the molar growth yield corrected for energy of maintenance. m_s is derived from a double reciprocal plot of the experimental values of Y_{glc} versus μ; the resulting straight line gives an intercept of $1/Y_{glc}^{max}$ and a slope of m_s. In the case of organisms whose fermentation pattern is affected by growth rate, these plots are not linear and in such instances it is necessary to plot Y_{ATP} versus μ. However, the original rather dubious assumption that the maintenance energy expenditure is independent of growth rate under all conditions has now been revised, largely as the result of investigations with aerobic carbon-limited and carbon-sufficient cultures; only at the maximum specific growth rate is the maintenance requirement identical under all growth conditions. These experiments showed that the carbon-sufficient cultures displayed uncoupling between energy-generation and utilization, leading to a larger proportion of the energy source being used for maintenance purposes, and that this varied with growth rate. Tempest introduced the term 'slip reactions' for these energy-spilling processes which can be accounted for by the existence ATPase activity, futile cycles, the formation of storage compounds, excretion of products of metabolism, e.g. pyruvate and acetate, and the presence of branched respiratory chains with non-phosphorylating segments (Tempest, 1978; Tempest and Neijssel, 1984). The excretion of acetate is particularly interesting because in its undissociated form acetate can function as a powerful uncoupler of oxidative phosphorylation; consequently at low pH values organisms excreting acetate will exhibit increased maintenance rates. Yield values for cultures displaying slip reactions would therefore be expected to be considerably lower than those computed on the basis of the energy requirements for polymer synthesis.

Pirt (1982) subsequently modified his original equation to take cognisance of the growth dependence of maintenance by adding to a constant maintenance energy term, m (independent of growth rate), a term m' which decreases linearly with increase in μ. Where q is the specific rate of energy source utilization, then the original equation becomes

$$q = \frac{\mu}{Y_{max}} + m$$

and this is modified to

$$q = \left(\frac{1}{Y_{max}} - \frac{m'}{\mu'_m} \right) \mu + m + m'$$

where μ'_m is the maximum specific growth rate when $m > 0$.

4.4.2 Energization of the cytoplasmic membrane

As we have seen (p. 60), the energized membrane plays a central role in driving ATP synthesis, transport processes and transhydrogenase reactions. Under aerobic conditions electron transport provides the energization directly without the intermediacy of ATP but anaerobically ATP is needed for this purpose. This point is admirably demonstrated by experiments with ATPase-negative mutants of *Escherichia coli* which, on account of their enzymic deficiency, cannot use ATP generated by substrate-level phosphorylation to produce a membrane potential and thus are unable to grow anaerobically (unless an acceptor such as nitrate or fumarate is present). These mutants can still, however, use respiration to energize transport, transhydrogenase and motility. Comparison of the growth characteristics of an ATPase-negative mutant with wild-type *E. coli* enabled Stouthamer to estimate that in anaerobiosis some 51% of total energy production is used for membrane energization. The assumptions made were that, under the experimental conditions used, the maintenance energy requirements of the two organisms are not disparate and their efficiency of biomass formation is identical. Then the difference in Y_{glc}^{max} for aerobic growth of the mutant (43.8) and anaerobic growth of the wild-type (20.8) will be a measure of the amount of energy needed for membrane energization of the wild type under anaerobic conditions.

The high energy expenditure for membrane energization is probably explained in part by the leakiness of the membrane both to protons and to intracellular metabolites. Although an essential postulate of the chemiosmotic hypothesis is the impermeability of the membrane to protons, experimental measurements of $\rightarrow H^+/O$ quotients reveal various degrees of leakiness. Consequently the membrane potential will be lowered by proton leakage, necessitating a higher energy expenditure to maintain the membrane in its energized state. For example, the ATPase-negative mutants of *E. coli* display a lower permeability to protons after anaerobic growth with nitrate as the terminal electron acceptor than after aerobic growth, indicating that some regulation of proton conduction is exerted by the cell, presumably to maintain the Δp (p. 61) at an appropriate value; this has obvious implications for energy expenditure.

Although motility and chemotaxis are dependent upon an energized membrane state, present evidence suggests that these processes do not make high energy demands on the cell.

CHAPTER FIVE

ENERGY TRANSDUCTION IN MEMBRANES

One of the crucial advances in the study of bioenergetics was the discovery that the major proportion of the ATP synthesized in cells is derived from membrane-bound enzyme complexes associated with the inner membrane of mitochondria, the thylakoid membrane of chloroplasts, and the cytoplasmic membrane of prokaryotes (bacteria and cyanobacteria). These membranes, referred to as *energy-transducing* or *coupling membranes*, are characterized by the possession of two types of protein assembly. The first is ATP phosphohydrolase (ATPase, or preferably ATP synthetase, since its principal function is the synthesis of ATP from ADP and P_i); this enzyme complex is found in all such membranes. The nature of the second assembly is determined by the primary energy source for the membrane. Thus in mitochondria and bacteria it is a respiratory chain catalysing the transfer of reducing equivalents from substrates to oxygen or other suitable acceptor, whereas in chloroplasts and photosynthetic bacteria it is a system containing light-harvesting (antenna) pigments (chlorophyll or bacteriochlorophyll). The mechanism(s) whereby the two assemblies are coupled to permit energy transduction to occur was for many years an extremely vexed problem but the proposal advanced by Mitchell, that a proton gradient across the membrane provides the necessary means, has been subjected to rigorous investigation and now finds fairly general acceptance; it is a concept central to his unifying *chemiosmotic hypothesis* (Mitchell, 1966) which, in offering an explanation for membrane energization, can also account for reversed electron transfer, the transport of certain solutes, cell motility, and the synthesis of inorganic pyrophosphate by some photosynthetic bacteria.

5.1 The chemiosmotic hypothesis

The chemiosmotic hypothesis postulates that the enzymes and electron carriers in the energy-transducing membrane associated with phosphorylation are asymmetrically oriented so that they catalyse vectorial reactions which effect the transport of molecules, chemical groups and ions across the membrane. Some of these reactions are electrogenic, i.e. they result in the separation of electric charges within and across the membrane, the recombination of such charges being associated with the performance of chemical, osmotic and mechanical work. The essential features of the hypothesis are that (1) electron transfer results in the extrusion of protons to the outside of the membrane: (2) the membrane is impermeable to ions, expecially H^+ and OH^-, except via specific exchange-diffusion systems, and (3) within the membrane there is a proton-translocating ATP phosphohydrolase (ATP synthetase) which catalyses the reaction

$$H^+ + ADP^{3-} + HPO_4^{2-} + nH_{outside}^+ \rightleftharpoons nH_{inside}^+ + ATP^{4-} + H_2O$$

where n is the number of g-ions of H^+ translocated per mole of ATP synthesized and is expressed as the $\rightarrow H^+/ATP$ quotient. The other proton in the equation is involved in ionization changes in the reaction, i.e. it is scalar rather than vectorial. Since this reaction is reversible the accumulation of protons on the outside of the membrane will result in the synthesis of ATP with a concomitant removal of protons to the inside of the membrane (Figure 5.1). The net result of either electron transfer or the hydrolysis of ATP is thus the generation of gradients of (1) electrical or membrane potential ($\Delta\psi$) and (2) chemical potential, ΔpH (pH_{out} minus pH_{in}), with energy transduction occurring via a proton current (proticity) circulating through an insulating membrane and the adjoining bulk aqueous phases. These parameters give rise to a *protonmotive force* (Δp) defined by the equation

$$\Delta p = \Delta\psi - Z\Delta pH$$

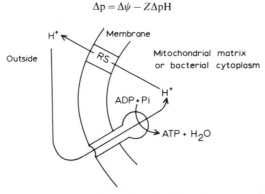

Figure 5.1. The proton current, continuously replenishing the protonmotive force, according to the chemiosmotic hypothesis. RS, redox system.

where Δp is in mV and $Z = 2.303\,RT/F$ (with a value of 60 mV at 30 °C). Consequently energy storage is a transmembrane phenomenon.

The protonmotive force which, according to the magnitude of the individual terms, may be a function primarily of $\Delta\psi$ or ΔpH, or a combination of the two, thus drives various transmembrane energy-dependent processes, including ATP synthesis by reversal of the proton-translocating ATP phosphohydrolase and the accumulation of metabolites via their respective transport systems. The physiological driving force for ATP synthesis differs according to the organelle, principally because of differences in the buffering capacity of the cytoplasm or medium and in the permeability of the membranes to ions. For example, $\Delta\psi$ is the dominant parameter in mitochondria whereas ΔpH predominates in chloroplasts. Bacteria utilize either $\Delta\psi$ or ΔpH or both depending upon environmental conditions (Table 5.1).

Table 5.1. Measurement of membrane potential ($\Delta\psi$) and protonmotive force (Δp) for mitochondria, chloroplasts and bacteria

Organelle or organism	pH_0	$-\Delta pH$	$-Z\Delta pH$ (mV)	$\Delta\psi$ (mV)	Δp (mV)
Mitochondria	6.5	0.81	47	137	184
	7.0	0.68	49	150	189
	7.6	0.47	27	135	162
Submitochondrial particles (beef heart)	7.5	1.8	95	90	185
Chloroplasts (thylakoids)	8.0	−3.0	−177	−100	−277
Escherichia coli	6.45	1.7	100	137	237
Streptococcus lactis	7.0	0.75	44.5	35.5	80
Staphylococcus aureus	6.65	1.15	68	128	196
Staphylococcus epidermidis	6.8	0	0	127	127
Micrococcus lysodeikticus					
Cells	5.5	1.4	86	110	196
Vesicles	5.5	1.4	86	82	168
Vesicles	8.0	0	0	107	107
Bacillus acidocaldarius	3.5	2.4	153	−36	117
Thermoplasma acidophilum	2.0	4.5	296	−120	176
Thiobacillus ferro-oxidans	2.0	4.5	266	−10	256
	3.0	3.3	192	40	232

Values are derived from the relationship $\Delta p = \Delta\psi - 2.303 RT/F\ \Delta pH$. pH_0 is the pH of the suspending medium and $\Delta pH = pH_0 - pH_i$ where pH_i is the pH of the matrix of the organelle or cytoplasm of the prokaryote. The Mitchell convention, that $\Delta\psi$ is positive when the matrix of the organelle or cytoplasm of the prokaryote is negative with respect to the suspending medium, is used. Apparent inconsistencies in the relationship of ΔpH(mV) in the table are attributable to differences in the temperatures at which the experiments were conducted as reflected by the value of the coefficient RT/F.

Note the relatively small contribution that $\Delta\psi$ makes to Δp in the acidophilic bacteria (bottom three organisms in the table).

Data have been collected from various sources; references to many original papers may be found in the review of Cobley and Cox (1983) (*Microbiol. Revs.* **47**, 579–595).

Certain consequences follow, given the validity of the chemiosmotic hypothesis. First, to maintain Δp there must be two compartments and gradients of $\Delta\psi$ and ΔpH can only be measured with membranes that form topologically closed units. Further, since Δp is inversely related to the rate of proton conductance through the membrane, agents which increase this rate effectively short-circuit the system by dissipating Δp while simultaneously stimulating the respiratory rate. Energy transduction is thus inhibited by such compounds which function as uncouplers (see below). Tight coupling in energy transduction implies that the rates of respiration and photosynthetic electron transfer can be controlled by back pressure of Δp, thereby ensuring cellular efficiency. Finally, uptake of protons with accompanying anions (symport) or in exchange for cations (antiport) is important for the transport of solutes across membranes via energy-dependent systems (p. 73).

The chemiosmotic hypothesis predicts that it should be possible to synthesize ATP by application of an artificially-generated Δp to a topologically intact membrane system which possesses ATP synthetase. This has, in fact, been demonstrated with chloroplasts, mitochondria and bacterial vesicles. Thus chloroplasts in the dark can be induced to synthesize ATP if their external pH is suddenly increased from 4 to 8, producing a transitory transmembrane ΔpH of 4 units; these organelles normally function with a high ΔpH and a low $\Delta\psi$. In contrast, mitochondria display opposite behaviour and the application of both a pH gradient and a K^+ diffusion potential permits ATP synthesis to occur.

There is now evidence to indicate that, in some organisms, if Δp falls below a critical value inhibition of the ATPase and of amino acid transport occurs, a phenomenon which is termed *gating*. This process, which serves to prevent ATP hydrolysis and to preserve intracellular metabolite pools, is possibly of significance for survival of bacteria under starvation conditions (p. 171).

The various processes of membrane energization and function are summarized schematically in Figure 5.2.

5.2 The structure and function of ATP synthetase

The ATP synthetase complexes of mitochondria, chloroplasts and bacteria are now known to be remarkably similar in structure (Futai and Karazawa, 1983). They consist of two multipeptide assemblies or coupling factors, designated F_0 and F_1 respectively (or BF_0 and BF_1 in the case of bacteria).

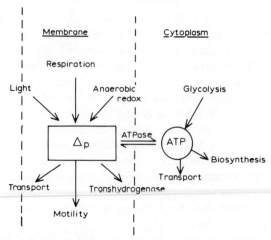

Figure 5.2. Schematic representation of membrane energization and function. After Harold (1978).

F_0 is hydrophobic and an intrinsic part of the membrane, providing the pore whereby the external protons gain access to the catalytic F_1 complex which, in contrast, is hydrophilic and situated on the internal side of the membrane where it protrudes as a knob-like structure. The F_1 complex can be detached from the membrane by treatment with urea or chelating agents or by lowering the ionic strength but then catalyses only ATP hydrolysis. It is possible that different catalytic sites exist on F_1 for ATP synthesis and hydrolysis. Associated with F_1 is a regulatory protein (I) whose principal function appears to be the prevention of hydrolysis of newly synthesized ATP. Although precise details of the mechanism await elucidation, ADP and P_i are believed to combine *in vivo* with the active site of F_1 with concomitant release of $2H^+$ to the internal compartment. Interestingly, unlike other ATPases, a covalent phosphoenzyme intermediate does not seem to be involved. The entry of $2H^+$ from the exterior via F_0 is associated with the reaction to form ATP and H_2O, which are released into the internal compartment. The possibility of membrane energization causing conformational changes in F_1 and the involvement of an alternating catalytic site mechanism have been proposed by Boyer (1977).

The contributions of Gibson and his colleagues to our understanding of oxidative phosphorylation have been seminal (for a review, see Gibson, 1982). These workers isolated mutants of *Escherichia coli* variously defective in respiratory chain electron transfer or in ATP synthesis, respectively due to their inability to synthesize ubiquinone (ubi^-) or the F_0 or F_1 components of ATP synthetase, the latter defect rendering them uncoupled (unc^-). The application of genetic techniques has proved particularly fruitful in structural and functional studies of bacterial ATP synthetases and the genes for each complex have been identified and mapped. The F_1 assembly consists of five different polypeptides, designated α, β, γ, δ and ε. A stoichiometry of $\alpha_3\beta_3\gamma\delta\varepsilon$ has been determined for the F_1 assemblies of $E.\ coli$ and the thermophilic bacterium PS3. The α and β subunits both have binding sites for ATP and ADP and are involved in the catalytic mechanism, δ and ε are required for the binding of F_1 to F_0, while the γ subunit links the $\alpha\beta$ catalytic moiety to the F_0-$\delta\varepsilon$ junction and is believed to function as a gate to proton conductance via the F_0 channel or pore in the coupling membrane (Figure 5.3).

The F_0 assembly in $E.\ coli$ comprises three polypeptides, designated a, b and c, but the stoichiometry remains controversial. Thus structures of $a_1b_2c_{10}$ and $a_1b_2c_{12-15}$ have been proposed; some bacterial F_0 assemblies lack subunit a and, indeed, that of $Clostridium\ pasteurianum$ has been claimed to consist solely of eight c subunits, which must therefore fulfil all the functions of F_0. Subunit c, which is common to all F_0 assemblies,

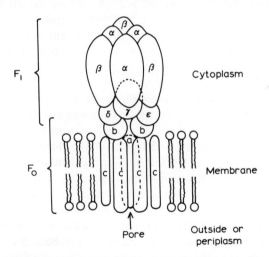

Figure 5.3. Model of subunit organization in F_0F_1 of $Escherichia\ coli$ ATP synthetase (not to scale). The F_0 assembly, located in the membrane and comprising subunits a, b and c, forms a proton channel (pore) through the membrane leading to assembly F_1 which protrudes into the cytoplasm. F_1 possesses catalytic subunits ($\alpha_3\beta_3$) bound to the F_0 assembly by subunits $\delta\varepsilon$. Subunit γ of F_1 is believed to serve as a 'gate' to the proton channel. Note that subunit a (F_0) is not present in all bacteria.

is an extremely hydrophobic proteolipid that binds N, N'-dicyclohexyl-carbodiimide (DCCD). It has a molecular weight of about 8k and forms the transmembrane pore or channel which permits protons and water to pass to and from the F_1 assembly. Chemical and genetic evidence suggests that the c subunits function as an oligomer. Subunit c of yeast and chloroplasts (but not of bacteria) has been found to translocate H^+ when reconstituted into liposomes.

The binding of DCCD to subunit c alters the proton permeability of F_0 and inhibits ATP synthesis whether mediated by respiration, light or an artificially imposed Δp. It also inhibits the generation of Δp via ATP hydrolysis. Conversely, DCCD can restore lost respiration-linked energy transduction stemming from increased proton permeability which is typical of membrane vesicles prepared from certain mutants of *E. coli* (*unc C*) uncoupled for oxidative phosphorylation. This pattern of behaviour supports the role of subunit c in forming the transmembrane proton channel.

Subunit b is a hydrophilic protein of molecular weight about 19k but which possesses a hydrophobic sequence of amino acid residues near its N-terminus by which embedding in the membrane occurs. As the extrinsic hydrophilic region of b can be digested with proteases without affecting the proton permeability of the F_0 assembly, it is deduced that the portion extruding from the membrane does not form the proton channel.

5.3 Uncouplers of oxidative phosphorylation

Several structurally dissimilar compounds are able to prevent the phosphorylation of ADP while simultaneously stimulating the respiratory rate of certain micro-organisms; they are referred to as *uncouplers*. The feature common to these compounds, and which holds the key to their action, is that each of them is a weak acid. If the uncoupler is lipid-soluble in both its protonated and unprotonated forms, then either molecule can traverse the membrane and, in so doing, effectively 'short-circuit' the electrochemical gradient, thereby enabling respiration to occur without stoichiometric ATP synthesis. For example, where a proton gradient exists (inside negative), the uncoupler will become protonated outside the cell and then cross the membrane to the interior where it will be deprotonated because of the alkaline environment. The deprotonated acid will then return to the exterior, completing the cycle, thus effectively dissipating the electrochemical gradient by functioning as an ionophore for protons. Examples of uncouplers include 2, 4-dinitrophenol (DNP), tetrachlorosalicylanilide

(TCS), carbonyl cyanide *m*-chlorophenyl hydrazone (CCCP) and carbonyl cynanide *p*-trifluoromethoxyphenylhydrazone (FCCP). The stimulatory effect of uncouplers on respiration is attributed to their relief of the mass action inhibition of the electron transfer chain by the electrochemical gradient it generates. By dissipating the proton gradient uncouplers promote electron flow and proton extrusion. They do not affect facilitated diffusion.

5.4 Determination of the protonmotive force

All methods for the determination of Δp require the separate measurement of ΔpH and $\Delta\psi$. ΔpH is usually calculated from the distribution at equilibrium of electrically neutral permeant weak acids and bases. $\Delta\psi$ is obtained by direct determination of the concentration gradient at equilibrium of an ion I^{m+} permeable by electrical uniport, and then applying the Nernst equation

$$\Delta\psi = 2.303 \frac{RT}{mF} \log \frac{[I^{m+}]_i}{[I^{m+}]_o}$$

where subscripts 'i' and 'o' indicate inside and outside respectively. Alternatively $\Delta\psi$ may be obtained spectroscopically by measuring the optical effect on an added probe indicator dye, such as 1-anilino-8-naphthalene sulphonate (ANS), of the diffusion potential generated by an ion gradient; the method is subject to uncertainties especially since factors other than $\Delta\psi$ or ΔpH may interfere (Bashford and Smith, 1979). Membrane-bound carotenoids of photosynthetic bacteria serve as valuable intrinsic probes of $\Delta\psi$ because they respond extremely rapidly to the electrogenic events initiated by the light reaction. However, a possible limitation is that their response is localized within the membrane and might not be truly representative of the bulk-phase membrane potential difference. General principles of measurement have been reviewed by Ferguson and Sorgato (1982). Experimental details for the measurement of ΔpH and $\Delta\psi$ are outside the scope of the present work and the reader is referred to the reviews of Bashford and Smith (1979), Ramos *et al.* (1979), Rottenberg (1979) and Skulachev (1979) for such information. Here a brief mention of the principles must suffice.

Microbial cells, with the exception of some fungi, are too small to permit the insertion of micro-electrodes, and thus to determine the concentration gradient of protons across the membrane (ΔpH) measurements are made of (1) the movement of H^+ into and out of cells or vesicles by determining the pH of the bulk medium or (2) the distribution of non-metabolized isotopically labelled organic acids or bases between the cells or vesicles and the external medium. It is also necessary to know the effective *intracellular* or *intravesicular volume*, which is achieved by determining the total water space (i.e. internal and external) of centrifuged or filtered cells and deducting from it the excluded water volume, i.e. the volume of water outside the cells. The former value is obtained by using tritiated water and the latter with a ^{14}C-labelled impermeant compound. More recently it has proved possible to determine the intracellular pH of micro-organisms by ^{31}P nuclear magnetic resonance measurements (Ugurbil *et al.*, 1979; Nicolay *et al.*, 1981). This non-invasive method depends upon the fact that the ^{31}P NMR chemical shift of intracellular orthophosphate is a function of the pH of the cell. At present the principal difficulty is the relatively low sensitivity of the technique, necessitating the use of dense microbial suspensions.

5.5 Motility and taxis

Motile micro-organisms display behavioural responses which operate principally to guide them away from locations in the environment that cannot support optimal energy generation and growth and towards those that can do so. Microbial movement occurs in response to such stimuli as oxygen, light or substrates (aerotaxis, phototaxis and chemotaxis, respectively) and depends upon sensory transduction and energization of the flagellar motors. Swimming is the result of a counterclockwise rotation of the flagella. It has been observed that bacteria in an isotropic environment display a random motility pattern consisting of straight 'runs' for about one second followed by 'tumbles', a kind of head-over-tail movement caused by reversal of the direction of rotation of the flagella; the organism then resumes swimming in a different, random direction.

The prokaryotic flagellum is a relatively simple structure composed of a filament, a hook and a basal structure which is attached to the bacterial cell envelope by a series of rings through which the filament passes (Simon *et al.*, 1978). There are two rings (S and M) in Gram-positive and four rings (L, P, S and M) in Gram-negative organisms. The semi-rigid filament is a long helical organelle containing 11 parallel strands of the protein flagellin. The M ring, to which the basal structure is secured, is rotated by the flagellar motor in a manner not yet fully understood. However, it has been established that a protonmotive force plays an important role in sensory transduction in response to both oxygen and light but not in response to solutes, the signals from which are conducted across the cytoplasmic membrane by three transmembrane-signalling proteins and converge at a common regulator for the flagellar motors. These signalling proteins apparently function by methylation or demethylation in response to attractants or repellants respectively. The principal factor regulating chemotaxis is the concentration gradient of the chemoeffector; a cell which senses an increase in attractant, or a decrease in repellant, represses tumbling and continues to swim in the favourable direction, while an increase in repellant enhances the probability of tumbling.

Aerotaxis is a universal property of obligate and facultative aerobes and evidence suggests that the terminal cytochrome oxidase of the respiratory chain is the oxygen receptor, e.g. cytochrome o in *Escherichia coli* and *Salmonella typhimurium*, and cytochrome aa_3 in *Bacillus cereus*. The bacteria swim up an oxygen gradient until the terminal oxidase is saturated with oxygen, but if they swim too far up the gradient they are then repelled by the high oxygen concentrations. Obligate anaerobes are repelled by oxygen although the receptors have not yet been identified. Thus oxygen functions both as an attractant and a repellant, and bacteria migrate to an environment which furnishes an oxygen concentration optimal to their mode of life (Taylor, 1983b). Because ATPase-negative mutants of *E. coli* display normal aerotaxis it is believed that this process is mediated by the protonmotive force and not by ATP (Taylor, 1983a).

Phototaxis requires not only the chlorophylls and accessory pigments but also the photosynthetic reaction centre (p. 133); mutants of *Rhodopseudomonas* spp. possessing photopigments but lacking the reaction centre do not display phototaxis. Again, it is believed that a protonmotive force mediates the process. The corollary to these observations is that any reagent which affects the chemiosmotic gradient should influence aerotaxis and phototaxis, a proposition substantiated by the use of uncouplers and respiratory inhibitors which dissipate the Δp and act as repellants. Energy demands on the microbial cell for motility are not considered to be very great. In *Paramecium* and other unicellular eukaryotes, an influx of Ca^{2+} produces an action potential in chemotaxis excitation. A mechanism of this type has never been detected in bacteria.

5.6 Bioluminescence

Luminous bacteria are found in marine environments throughout the world. Formerly placed in the genera *Rhodobacterium* and *Beneckea*, they

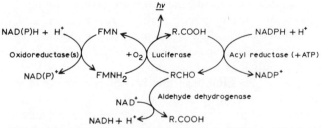

Figure 5.4. The sequence of reactions postulated to be involved in bacterial bioluminescence. The NAD(P)H:FMN oxidoreductases utilize soluble FMN which on reduction becomes a substrate for the luciferase reaction. After Ziegler and Baldwin (1981).

are now assigned to the genus *Vibrio*. These organisms possess the enzyme luciferase which operates as a mixed-function oxygenase and catalyses *in vitro* the light-generating reaction

$$FMNH_2 + RCHO + O_2 \rightarrow FMN + RCOOH + H_2O + hv$$

where RCHO is a long-chain aldehyde. The quantum yield of the bioluminescent reaction, defined as the total number of photons produced per molecule of substrate utilized or enzyme turned over, is recorded in units of einsteins mol^{-1}. Experiments *in vitro* suggest a quantum yield of about 0.1.

As the substrates for the reaction, namely $FMNH_2$ and the long-chain aldehyde, each require one molecule of NAD(P)H for their formation, the luminescent system has been regarded as a branch of the respiratory chain at the level of flavin (Figure 5.4). The diversion of electrons to the generation of light is energetically expensive for the cell and it has been estimated that some 6 ATP molecules are effectively lost per luciferase turnover. Since very active luminous bacteria can emit over 10^4 photons sec^{-1} per cell, if the *in-vitro* quantum yield of 0.1 is also assumed for *in-vivo* conditions (it is almost certainly an underestimate), then the expenditure per photon could be as high as 60 ATP molecules. Calculations have been carried out which indicate that, depending upon the luminescent activity of the bacterium, energy expended on light generation may represent from about 20% to as much as 200% of the energy required for growth. It seems reasonable to suppose that such an energetically costly system must subserve some important function for it to have persisted throughout evolution; presumably light emission confers a selective advantage, permitting the perception of luminous bacteria by other organisms with which they can establish a symbiotic relationship, e.g. as gut symbionts.

CHAPTER SIX

TRANSPORT PROCESSES

The growth and survival of a micro-organism depends upon its ability to transport essential nutrients from the environment across the membrane and into its cytoplasm. The molecules involved may be electrically charged or neutral and their uptake by the cell can be either passive or active, the latter requiring the expenditure of energy. The energy-independent passive processes are of two types, namely (1) simple diffusion and (2) facilitated diffusion, which is a carrier-mediated process and thus displays saturation kinetics. Both processes usually result in equilibration of a solute across the membrane, i.e. $[S]_i = [S]_o$, where subscripts 'i' and 'o' refer to inside and outside respectively, unless S is charged, in which case if there is an independently maintained membrane potential ψ, $[S]_i \neq [S]_o$. The general relationship is

$$\Delta G = 2.303RT \log \frac{[S]_i}{[S]_o} - mF\Delta\psi \qquad (6.1)$$

where m is the net charge on S and F is the Faraday constant. At equilibrium $\Delta G = 0$ and thus if the term $mF\Delta\psi$ is finite then distribution of S on either side of the membrane will be unequal. However, the concentration gradients achieved in this way are never very high.

In contrast, active transport, involving energy expenditure, enables a microbial cell to concentrate solutes (without any chemical modification) to high internal concentrations, i.e. the solute is transported against the concentration gradient. The process of uptake against a concentration gradient is driven by being coupled to metabolic reactions such that the net process has a negative ΔG. A specialized form of active transport is group translocation in which the solute undergoes chemical modification (usually phosphorylation) as it traverses the membrane. Active transport is, of course, sensitive to the action of uncoupling agents which dissipate the electrochemical gradient of the membrane.

6.1 Conceptual models of transport: ionophores

There are two conceptual models of the way in which a solute molecule traverses a membrane. The first is the *mobile carrier* mechanism which considers that the solute-binding moiety of the transport system alters its orientation, either by rotating across the plane of the membrane or by serving as a shuttle between outer and inner surfaces. The second is the *pore* model which assumes that the carrier protein remains essentially static in the membrane and forms a hydrophilic channel or pore specific for a given solute. It is believed to have a gate mechanism so that the pore permits the passage only of its specific solute.

Evidence for both of these mechanisms has been derived from experiments with the antibiotics valinomycin and gramicidin A. Valinomycin is a cyclic depsipeptide in which the hydrophilic groups of the constituent amino-acid residues are directed internally while the hydrophobic groups are exposed to the solvent. It is highly specific in binding the alkali metals K^+, Rb^+ or Cs^+ (but not Li^+ and Na^+) within its annulus and with a stoichiometry of one ion per valinomycin molecule. The resulting valinomycin complex enables the water-soluble ion to be carried across the lipid bilayer membrane shielded within the lipid-soluble antibiotic structure. Its ability to do so depends upon the physical state of the membrane, as shown by temperature-dependence experiments; valinomycin-K^+ transport increased markedly above the phase-transition temperature of an artificial phospholipid bilayer indicating that decreased viscosity facilitated the rate of uptake. Valinomycin thus functions as a mobile carrier that can diffuse through the membrane and as such is widely used in experimental systems.

Gramicidin A is a linear peptide composed of 15 amino acids: it binds cations (with a lower specificity than valinomycin) and transports them across membranes. In contrast to valinomycin, however, K^+ transport by gramicidin A is independent of temperature, suggesting that its effectiveness is not related to the viscosity of the membrane. It is concluded, therefore, that gramicidin A functions as a static pore in the membrane whereas valinomycin serves as a mobile carrier. The pore structure is probably formed by head-to-head dimerization of helical gramicidin A monomers, producing a hydrophilic aqueous channel along the axis of the helix. Other ionophores include 2, 4-dinitrophenol, which serves as a mobile proton carrier, and nigericin, which catalyses the exchange of H^+ and K^+ across membranes with a 1:1 stoichiometry and also operates in a mobile fashion.

6.2 Active transport: primary and secondary systems

It is possible to distinguish between primary and secondary active transport systems. The former comprise those mechanisms which convert radiant or chemical energy into electro-osmotic energy, i.e. the photophosphorylation system, the proton-translocating ATPase and the respiratory chain, all of which generate proton gradients. Secondary active transport systems link the flux of compounds such as certain sugars, amino acids and some ions to the flux of a substrate for a primary transport system, principally H^+ or Na^+. Consequently the movement of these ions *down* their electrochemical gradient is coupled to the movement of the co-transported molecule *up* its electrochemical gradient.

Three types of secondary co-transport are recognized and Mitchell has designated them as symport, antiport and uniport, the carrier molecule being bifunctional (Figure 6.1). Symport occurs when two different molecules or ions are simultaneously transported by the same carrier in the same direction. Antiport is similar except that the two different solutes are transported in opposite directions, while uniport involves a carrier which ferries only one solute. In the case of symport and antiport the uptake may be electroneutral or electrogenic depending upon whether the proton translocation is accompanied by the movement of a compensating anion, or by a neutral solute. In uniport cations move into and anions out of the cell in response to the membrane potential.

For coupled transport to occur the free energy required to transport solute S against an electrochemical gradient is given by equation 6.1. This energy must be provided by the Δp generated by the primary active transport system, and since $\Delta G = nF\Delta p$, then

$$\Delta G = nF\Delta p = 2.303RT\log\frac{[S]_i}{[S]_o} - mF\Delta\psi \qquad (6.2)$$

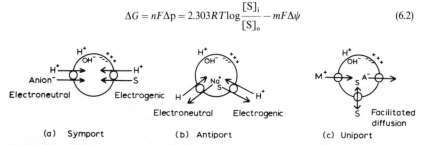

Figure 6.1. Schematic representation of secondary active transport mechanisms. In symport (*a*) and antiport (*b*), 1:1 co-transport with a proton occurs. In uniport (*c*), cations move into and anions out of the cell in response to the membrane potential while the passive movement of the uncharged solute S is facilitated by the carrier. After Dawes (1980).

where n is the number of g-ions of protons linked with the transport of 1 mole of S. Equation 6.2 can be written as

$$n\Delta\psi - nZpH = Z \log\frac{[S]_i}{[S]_o} - m\Delta\psi \qquad (6.3)$$

where $Z = 2.303RT/F$. By rearranging,

$$\log\frac{[S]_i}{[S]_o} = \frac{(n+m)\Delta\psi - nZ\Delta pH}{Z} \qquad (6.4)$$

and the concentration ratio of the solute inside to outside of the cell which can be supported by a given Δp can be ascertained. Equation 6.4 is valid for the three types of co-transport. Thus, in symport, if a proton and a monovalent anion are co-transported in the ratio 1:1 (i.e. electroneutral), equation 6.4 becomes

$$\log\frac{[S]_i}{[S]_o} = -\Delta pH \qquad (6.5)$$

which indicates that the concentration ratio of the anion inside to outside is sustained by a driving force equivalent to the pH gradient alone. Conversely, in antiport, where co-transport occurs in opposite directions, a proton–cation system in which both m and n are unity is equivalent to a hydroxyl–cation symport where $n = -1$. Thus for such an antiport system equation 6.5 holds with sign reversed, i.e. $\log [S]_i/[S]_o = \Delta pH$.

Should symport involve an uncharged solute in the ratio of 1:1 with a proton then the process becomes electrogenic, because each translocation event accumulates one positive charge within the cell. Consequently, equation 6.5 becomes

$$\log\frac{[S]_i}{[S]_o} = \frac{\Delta\psi}{Z} - \Delta pH = \frac{\Delta p}{Z} \qquad (6.6)$$

Thus electrogenic symport processes are driven by the total protonmotive force, in contrast with electroneutral ones which employ the pH gradient alone. Obviously the situation becomes more complex when n and m differ from unity.

Uniport processes have $n = 0$, thus equation 6.4 becomes

$$\log\frac{[S]_i}{[S]_o} = \frac{m\Delta\psi}{Z} \qquad (6.7)$$

which is essentially a form of the Nernst equation. If the solute is uncharged, m is zero and the uniport is a facilitated diffusion mechanism with a carrier not coupled to a potential gradient and consequently unable to concentrate the solute, which is simply equilibrated across the membrane.

6.3 Group translocation

The essential feature of group translocation is that the substrate transported arrives inside the cell in a chemically modified form, usually as its phosphate ester. The phosphoenolpyruvate (PEP) phosphotransferase (PT) system mediates the uptake of many, but not all, carbohydrates and occurs principally in anaerobic and facultative bacteria, e.g. enteric bacteria and streptococci. There is no evidence for its presence in fungi. It comprises a cytoplasmic component (Enzyme I), a membrane-bound Enzyme II that is usually specific for one sugar, Enzyme III, also sugar-specific, which is required for the transport of some but not all sugars and which may, or may not, be membrane-bound, and a small protein, HPr ($M = 9.5$k). In a series of reactions (Fig. 6.2), the sugar taken up from the environment becomes phosphorylated in the 6-position as it traverses the membrane. Energy (equivalent to 1 ATP) is expended as the phosphoryl group of PEP is transferred sequentially via Enzyme I, HPr and Enzyme III to the integral membrane Enzyme II, which effects the vectorial phosphorylation of the sugar. The phosphate group is bound to histidine residues in Enzyme I and HPr and possibly also in Enzymes III and II.

The PT system is linked to the regulation of the activity of some secondary transport systems; this occurs by a direct inhibition of carrier proteins by the unphosphorylated form of Enzyme III and by a stimulatory effect of the phosphorylated form of this enzyme on adenylate cyclase, the

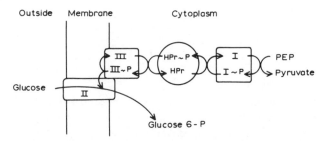

Figure 6.2. Schematic representation of the reactions effected by the phosphoenolpyruvate phosphotransferase system for sugar uptake. Enzymes I, II and III operate as shown but note that III is not always present and may or may not be membrane-bound.

enzyme which synthesizes cAMP, thereby increasing cAMP levels. cAMP binds to the receptor protein and promotes the synthesis of enzymes for the catabolism of non-PT system substrates. The overall regulation achieved is that in the presence of sugars transported by the PT system the synthesis of carrier proteins for certain non-PT substrates is repressed, while in their absence it is promoted.

Other group translocations in micro-organisms include the uptake of fatty acids in *Escherichia coli*, which appears to involve the acyl-SCoA derivatives within the cell, and the transport of purine and pyrimidine bases by enteric bacteria. Nucleosides are hydrolysed to their free bases which are then group translocated by a phosphoribosyltransferase with intracellular accumulation of the purine nucleoside monophosphate and inorganic pyrophosphate. However, there is a significant difference from the PT system because the phosphoribosyltransferase is a periplasmic enzyme easily lost by osmotic shock treatment of the bacteria.

Studies with vesicles prepared from bacterial membranes have demonstrated the presence of both active transport and group translocation mechanisms (Konings, 1977). Transport by vesicles possessing rhodopsin, isolated from the photosynthetic *Halobacterium halobium*, can be energized by light.

6.4 ATP-dependent transport systems

Gram-negative bacteria possess transport systems for certain amino acids and for phosphate which involve binding proteins located in the periplasmic space. The activities of these systems are lost by osmotic shock treatment; acetyl phosphate seems to furnish the direct energy source for their operation.

Both Gram-positive and Gram-negative bacteria also appear to have some ATP-dependent transport systems which do not involve periplasmic binding proteins, e.g. Na$^+$ transport in *Streptococcus faecalis*.

6.5 Solute transport in fungi

It has been established that the primary cation pump in fungi is an electrogenic, proton-translocating ATPase and that most of the carriers are proton-specific. In *Neurospora crassa* there is also a Ca^{2+}/H$^+$ antiport system with a presumed physiological function of Ca^{2+} extrusion. The H$^+$-ATPase functions by the enzyme being phosphorylated by ATP and hydrolytically dephosphorylated during its catalytic cycle (Scarborough, 1985).

Proton symport mechanisms have been identified in yeasts (*Saccharomyces* spp. and *Candida* spp.) for amino acids, phosphate and certain sugars. These exhibit proton: solute stoichiometries of 1:1 or 2:1, with the possibility of 3:1 in the case of glutamate and phosphate. Much remains to be learned about eukaryotic microbial transport, a subject which has been reviewed by Eddy (1982).

6.6 Mitochondrial solute transport

Compartmentation of the eukaryotic microbial cell presents problems of intracellular transport between the cytoplasm and the mitochondria. Mitochondria possess an outer membrane freely permeable to most solutes and an inner, invaginated, semi-permeable membrane, in which reside enzymes and electron carriers of the respiratory chain (Chapter 7). Because glycolysis and the pentose phosphate cycle occur in the cytoplasm whereas terminal oxidation via the tricarboxylic acid cycle and the β-oxidation of fatty acids take place within the mitochondria, transport of metabolites between cytoplasm and mitochondria and *vice versa* is of paramount importance.

While uncharged molecules are often able to permeate the inner membrane easily, charged molecules require carrier mechanisms, and systems for the translocation of phosphate, sulphate, sulphite, ADP, ATP, pyruvate and organic acids of the tricarboxylic acid cycle, glutamate, aspartate, proline, orthinine, citrulline, acyl carnitines and phosphoenolpyruvate have been identified. Some ions and metabolites do not readily permeate the inner mitochondrial membrane: they include $NAD(P)^{+}$, $NAD(P)H$, coenzyme A and its acyl esters, AMP, GMP, GDP, GTP, Cl^{-}, NO_3^{-}, Br^{-} and NH_4^{+}. On the other hand, water, oxygen, carbon dioxide, ammonia, formate, acetate, propionate and butyrate permeate freely without the necessity for a carrier system. Space does not permit an extended survey of mitochondrial transport systems and here attention will be confined to representative examples important for energy transduction; for further information the reader is directed to Prebble (1981).

Energy transduction via the process of oxidative phosphorylation (Chapter 7) requires the transport of inorganic phosphate and ADP into, and ATP out of, the matrix of the mitochondrion. Two mechanisms have been recognized for phosphate translocation on the basis of differential inhibition, namely P_i/OH^- exchange and P_i/dicarboxylate exchange diffusion systems. The former, representing electroneutral antiport $(H_2PO_4^-/OH^-)$, is promoted by the existence of a proton gradient (ΔpH; inside, i.e. matrix, negative) generated by the respiratory chain, although it is not, in fact, possible to distinguish between this exchange and entry of phosphate accompanied by a proton ($H_2PO_4^-/H^+$ symport). The phosphate-dicarboxylate exchange system was discovered when it was realized that succinate entry to the mitochondrion required the exit of phosphate. This carrier exchanges phosphate and succinate or malate, and phosphate transport may therefore be regarded as the primary event in a sequence since the exit of this anion drives the uptake of malate and malate, in turn, exchanges with tricarboxylate anions. Thus the exit of malate is linked to the uptake of citrate via a tricarboxylate carrier which plays an important role, for example, in lipogenic (oleaginous) yeasts. These organisms, which synthesize large amounts of lipid, need to transfer acetyl groups from the mitochondrial matrix, where they are formed, to the cytosol where lipid synthesis occurs. The membrane is impermeable to acetyl-SCoA but this substrate is reacted with oxaloacetate, catalysed by citrate synthase, and the citrate formed is transported to the cytosol by the tricarboxylate carrier. There the citrate is cleaved by citrate lyase to acetyl-SCoA and oxaloacetate.

Adenine nucleotide transport is effected by a translocase specific for ADP and ATP, which are exchanged across the membrane in a 1:1 ratio. As the two nucleotides are of different charge (ADP^{3-}, ATP^{4-}), the translocase effects electrogenic antiport in response to the membrane potential ($\Delta\psi$, negative inside), which increases the rates of ADP entry and ATP exit relative to those characteristic of the de-energized state. The factors influencing adenine nucleotide translocation are the adenine nucleotide requirements of the cytoplasm for biosynthesis (indicated by a high [ATP]/[ADP] ratio) and of the mitochondrial matrix for oxidative phosphorylation (a low [ATP]/[ADP] ratio).

Nicotinamide nucleotides do not penetrate the inner mitochondrial membrane but, aerobically, NADH generated by glycolysis in the cytoplasm must be oxidized via the respiratory chain. The difficulty is overcome by translocating reducing equivalents across the membrane indirectly via metabolic shuttles. The principal mammalian example is the *malate–aspartate shuttle* (Borst cycle) which involves intra- and extramitochondrial malate dehydrogenases and aminotransferases (transaminases) in association with the glutamate–aspartate and malate-2-oxoglutarate translocases (Figure 6.3); the net result is equivalent to the entry of NADH

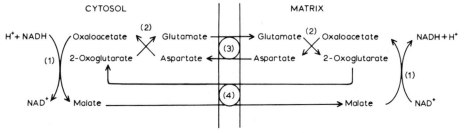

Figure 6.3. The malate–aspartate shuttle or Borst cycle which transports reducing equivalents across the inner mitochondrial membrane. Enzymes: (1) malate dehydrogenase; (2) aspartate aminotransferase; (3) glutamate-aspartate translocase; (4) malate-2-oxoglutarate (dicarboxylate) translocase.

into the mitochondrion. Operation of the shuttle, which is presumed to occur in microbial eukaryotes, does not expend ATP. Another system, the *glycerol 3-phosphate shuttle*, which is principally found in insect mitochondria, results in ATP loss because it involves cytoplasmic NAD^+-linked and mitochondrial FAD-linked glycerol 3-phosphate dehydrogenases; $FADH_2$ is generated at the expense of NADH with accompanying loss of opportunity for site I ATP synthesis. It has been proposed that this mechanism operates in yeasts, e.g. *Candida tropicalis*, growing on alkanes. The β-oxidation of fatty acids produced in alkane metabolism occurs cytoplasmically in peroxisomes, organelles that appear in cells grown under these conditions. NADH generated by 3-hydroxyacyl-SCoA dehydrogenase is reoxidized by an NAD^+-dependent glycerol 3-phosphate dehydrogenase effecting the reduction of dihydroxyacetone phosphate; the resulting glycerol 3-phosphate then participates in the shuttle, enters the mitochondrion and is oxidized by an FAD-specific dehydrogenase.

The growth of alkane-utilizing yeasts offers an interesting problem in bioenergetics for not only is there effective loss of ATP via the glycerol 3-phosphate shuttle in fatty acid oxidation but also in the peroxisomes. Thus $FADH_2$ generated by the action of acyl-SCoA dehydrogenase is reoxidized with the formation of H_2O_2, which is decomposed by the catalase present in these organelles, and therefore does not contribute to oxidative phosphorylation in the mitochondria. Consequently, whereas the customary mitochondrial β-oxidation of fatty acids yields 5ATP per turn of the fatty acid oxidation spiral (3ATP and 2ATP respectively from the oxidation of NADH and $FADH_2$), peroxisomal oxidation yields but 2ATP, via the glycerol 3-phosphate shuttle, i.e. it is only 40% as efficient (Boulton and Ratledge, 1984).

A possible explanation is offered by the fact that alkanes are highly reduced and may be considered to be 'energy rich, carbon limiting' compounds (Anthony, 1980), i.e. the oxidation of an alkane generates more ATP than is needed for the assimilation of all of its carbon and therefore energy not conserved as ATP is dissipated as heat via the peroxisomal catalase reaction. From this standpoint carbohydrates may be regarded as 'energy limited, carbon excess' compounds because glucose, being an oxidized compound, does not in its metabolism yield sufficient ATP to permit all the available carbon of the substrate to be assimilated, and some is therefore oxidized to CO_2 and water ('wasted') in simultaneously furnishing energy for the assimilation of the remainder of the molecule.

It must be borne in mind, too, that certain cytoplasmic metabolic processes, e.g. gluconeogenesis, require reducing equivalents from the mitochondrion in order to maintain the necessary $[NAD^+]/[NADH]$ ratios for their operation. Further, there is normally a significant difference in this ratio across the inner mitochondrial membrane, values of between 300 and 1000 being cited for the cytosol compared with about 10 for the mitochondrial matrix; other work indicates this difference in terms of redox potential to be $\simeq 50\,mV$, the maintenance of which requires energy expenditure of some $9.6\,kJ\,mol^{-1}$.

CHAPTER SEVEN

ENERGY CONSERVATION IN CHEMOHETERO-TROPHIC AEROBIC METABOLISM

The energy required by aerobic heterotrophic micro-organisms is principally derived from the processes of electron transfer and oxidative phosphorylation, representing the final stages in the flow of electrons from organic substrates to oxygen. Hydrogen atom pairs (equivalent to $2H^+ + 2e^-$) are removed by specific dehydrogenase enzymes from certain intermediates of carbohydrate, fat and protein catabolism, and together with those arising from the oxidative steps of the tricarboxylic acid cycle, are fed into the electron transport or respiratory chain, donating their electrons to carrier components of the chain while the accompanying H^+ ions enter the aqueous environment. The electrons are sequentially transferred along this chain of redox components of gradually increasing standard electrode potential[1] until they reach *cytochrome oxidase* which catalyses the transfer of electrons to oxygen, the terminal electron acceptor of aerobic organisms. Each atom of oxygen accepts two electrons from the chain and takes up $2H^+$ from the environment (equivalent to the $2H^+$ associated with the $2e^-$ which entered the chain) and water is formed. The electron carriers are assembled in a highly organized manner in the inner mitochondrial membrane of eukaryotes and in the cytoplasmic (plasma) membrane of prokaryotes.

Each component of the electron transport chain comprises a redox couple whose E_m displays a fairly small difference from its neighbours, consequently the transfer of electrons from donor to acceptor is associated with the release of free energy in relatively small amounts, which facilitates energy conservation. The free energy change (ΔG) of the reaction involving electron transfer from a donor redox couple to an acceptor couple is given

---

[1] Determination of the standard electrode potential (E'_0) of a membrane-bound redox component presents problems and the nearest equivalent that can be measured is recorded as E_m, the mid-point potential.

by the equation $\Delta G = -nF\Delta E$, where n is the number of electrons transferred, F is the Faraday constant (equal to 96 486 coulombs) and ΔE the difference in electrode potential between the two redox couples. Likewise, the standard free energy change of the redox reaction is related to the difference in mid-point potentials of the two redox couples by the expression $\Delta G^{0'} = -nF\Delta E_m$. Molecular mechanisms in the membrane effectively couple electron transport to the regeneration of ATP from ADP and P_i; for this reason the membranes are often referred to as *coupling membranes* (p. 60). The coupled synthesis of ATP occurs at certain locations in the respiratory chain called the *energy-conserving sites* or *segments*. In eukaryotic membranes there are three such sites, permitting 3ATP to be formed per pair of electrons transported to oxygen, i.e. a $P/2e^-$ ratio of 3 (Figure 7.1). While the respiratory chains of mammalian and yeast membranes are very similar, those of few bacteria resemble them. Indeed, according to species, bacteria exhibit a wide variety of respiratory chains the individual components of which are often affected by growth conditions. Generally there are at most two energy conservation sites in bacteria.

7.1 The components of the respiratory chain

Mitochondrial respiratory chains comprise a linear sequence of redox components consisting of flavoproteins and iron–sulphur (Fe–S) proteins (the dehydrogenases), quinones, cytochromes and cytochrome oxidase (cyt aa_3) (Figure 7.1). Most of the electrons are fed into the chain via NADH, the cofactor for the majority of the dehydrogenases involved in catabolism,

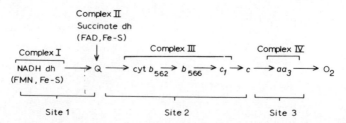

Figure 7.1. The eukaryotic respiratory chain located in the inner mitochondrial membrane. The four multiprotein assemblies are designated as Complexes I to IV and the three sites of coupling of electron transfer to ATP regeneration from ADP and P_i are indicated. Succinate oxidation feeds reducing equivalents directly to Q and by-passes Site 1 (NADH dh, NADH dehydrogenase; FMN. flavin mononucleotide: FAD. flavin adenine dinucleotide; Fe-S, non-haem iron-sulphur centre; Q, ubiquinone; cyt, cytochrome).

and which thus serves as a collecting agent. NADH is reoxidized by *NADH dehydrogenase*, a flavoprotein enzyme possessing a flavin mononucleotide (FMN) prosthetic group and which is associated with two other protein units containing respectively two and four Fe–S centres. The FMN is reduced to $FMNH_2$ and reoxidized by the iron atoms of the Fe–S centres which transfer reducing equivalents from it to ubiquinone, the next carrier in the chain, and undergo $Fe^{2+} - Fe^{3+}$ cycles in the process. The complex of NADH dehydrogenase with the Fe–S proteins is termed *NADH-ubiquinone reductase* or *complex I*. Succinate dehydrogenase differs from many other dehydrogenases in having flavin adenine dinucleotide (FAD) as prosthetic group and Fe–S centres; NAD^+ reduction is thus by-passed and electrons are fed directly to ubiquinone. There are two protein subunits, one containing one molecule of FAD and four Fe–S centres and the other an iron-sulphur protein with four Fe–S centres. These proteins require also cytochrome b_{558} and a small peptide unit to permit reduction of ubiquinone and are collectively referred to as *succinate–ubiquinone reductase* or *complex II*. Other FAD-enzymes which donate electrons directly to ubiquinone include fatty acyl-SCoA dehydrogenase and glycerol phosphate dehydrogenase.

Ubiquinone (coenzyme Q) is a fat-soluble quinone with a long side chain of isoprenoid (5-carbon) units. In ubiquinone from most mammalian tissues and yeasts there are ten units (Q_{10}), although in other organisms there may be only six or eight isoprenoid units. While Gram-negative bacteria also employ ubiquinone, Gram-positives generally possess *menaquinone* (MK, vitamin K_2) and facultative organisms like *Escherichia coli* have both, commonly using ubiquinone for aerobic respiration. (Ubiquinone and menaquinone possess substituted 1,4-benzoquinone and 1,4-naphthoquinone nuclei respectively. Related *demethylmenaquinones* (DMK), lacking the nuclear methyl group, are present in some species of *Streptococcus* and *Haemophilus*.) All the quinones function by accepting reducing equivalents from the NADH-ubiquinone reductase and from FAD-linked dehydrogenases, being reduced to the quinol form (QH_2).

The reoxidation of the quinol is effected by a series of red-brown proteins, the *cytochromes*, each of which contains an iron–porphyrin (haem) prosthetic group. In eukaryotes there are three classes of cytochrome arranged in the sequence $b_{562} \rightarrow b_{566} \rightarrow c_1 \rightarrow c \rightarrow aa_3$.

The central iron atom of each haem unit accepts one electron and undergoes an $Fe^{2+} - Fe^{3+}$ cycle; because the quinones donate two reducing equivalents, two haem units are required per quinone oxidized. There is a fourth type of cytochrome, designated *d*, present in bacteria and some protozoa which functions as a cytochrome oxidase and replaces aa_3, and also cytochrome o which does likewise.

The cytochromes differ principally in the substituent side chains of their porphyrin nucleus, and they display different absorption maxima within the range of about 550 to 650 nm, and different E_m values, spanning approximately -1000 to $+400$ mV. Individual cytochromes are usually distinguished by subscripts indicating their absorption maximum in nm. Cytochromes b_{562} and b_{566} are extremely hydrophobic and are tightly bound in the membrane whereas cytochrome c is very hydrophilic and can readily be extracted.

The transfer of electrons from ubiquinone to cytochrome c involves *ubiquinol cytochrome c reductase* or *complex III*, which comprises cytochromes b_{562}, b_{566} and c together with a protein ($M = 26k$) possessing two Fe–S centres. Reduced cytochrome c is then reoxidized by *cytochrome oxidase* (cytochrome aa_3 or *complex IV*), the terminal enzyme of the respiratory chain which catalyses the reaction

$$4H^+ + 4e^- + O_2 \rightarrow 2H_2O$$

accepting electrons from four molecules of ferro-cytochrome c and four protons from the aqueous environment. This complex comprises two different haems, a and a_3 with respective E_m values of $+245$ mV and $+340$ mV, only the latter of which reacts with oxygen, and two copper atoms (Cu_A and Cu_B) which undergo a $Cu^+ - Cu^{2+}$ cycle, each associated with one of the haems, e.g. $Cu_A^{2+} \cdot a^{3+} . Cu_B^{2+} a_3^{3+}$ represents its fully oxidized state.

As already noted, cytochrome d functions as an oxidase in many bacteria; it contains two haem d, having E_m of $+245$ mV and $+345$ mV, and two haem b. In the cytochrome d of protozoa, however, two copper atoms replace the two haem b units.

Many organisms possess a membrane-bound *nicotinamide nucleotide transhydrogenase* ($M = 94k$) which effects the reversible transfer of a hydride ion (H^-) between NADPH and NAD$^+$, giving an overall reaction

$$NADPH + H^+ + NAD^+ \rightleftharpoons NADP^+ + H^+ + NADH$$

The equilibrium constant of the reaction is close to unity but the equilibrium is shifted by linking the reaction with ATP utilization or the proton-motive force generated by respiration, thus favouring NADPH formation at the expense of NADH produced by the mainly NAD-linked dehydrogenases of catabolism. The resulting NADPH is then available for the NADP$^+$-linked dehydrogenases characteristic of anabolic sequences. The net effect, therefore, is that energy is utilized to direct reducing power from catabolism to biosynthetic purposes in a process termed *reversed electron transfer*. The converse situation of the transhydrogenase contributing to energy generation via the respiratory chain is very unlikely under physiological conditions because the extremely high ratio of [NADPH] [NAD$^+$]/[NADP$^+$] [NADH] that would be required is probably unattainable *in vivo*. Nonetheless it does represent a potential site of energy conservation and is referred to as site O.

Energy-independent transhydrogenases are also found in some bacteria, e.g. the genus *Pseudomonas*. They may be soluble or membrane-bound and often contain flavin. Their principal function is believed to be the transfer of any excess reducing equivalents derived from NADP$^+$-linked substrates to NAD$^+$ for oxidation via the respiratory chain.

In summary, the mitochondrial respiratory chain comprises four distinct complexes or assemblies that can be isolated from the inner membrane and are linked by redox carriers *in situ*. By suitable techniques (p. 92), it is possible to show that complexes I, III and IV correspond to energy conservation segments of the chain where coupling to ATP synthesis occurs, referred to respectively as sites 1, 2 and 3.

7.2 Bacterial respiratory chains

Bacteria do not possess mitochondria and their respiratory chains reside in the cytoplasmic membrane. While there is some superficial resemblance to the mammalian and yeast systems just surveyed, there are generally significant differences. There are also differences between bacterial species themselves (Haddock and Jones, 1977; Ingledew and Poole, 1984); these we shall now consider. Nonetheless, a few bacteria do display similarity to mitochondrial-type respiratory chains both in respect of their electron carriers and in their sensitivity to inhibitors of electron transport, e.g. *Paracoccus denitrificans* (Figure 7.2) and *Alcaligenes eutrophus*. John and Whatley (1975) have, in fact, proposed that the inner membrane of the contemporary mitochondrion probably evolved from an ancestral relative of *P. denitrificans* via endosymbiosis with a primitive host cell.

The vast majority of heterotrophic bacteria exhibit important differences from the mitochondrial respiratory system in four respects.

(1) The replacement of one redox carrier by another which possesses similar properties. We have already noted that menaquinone (MK) or demethylmenaquinone (DMK) replaces ubiquinone in most Gram-

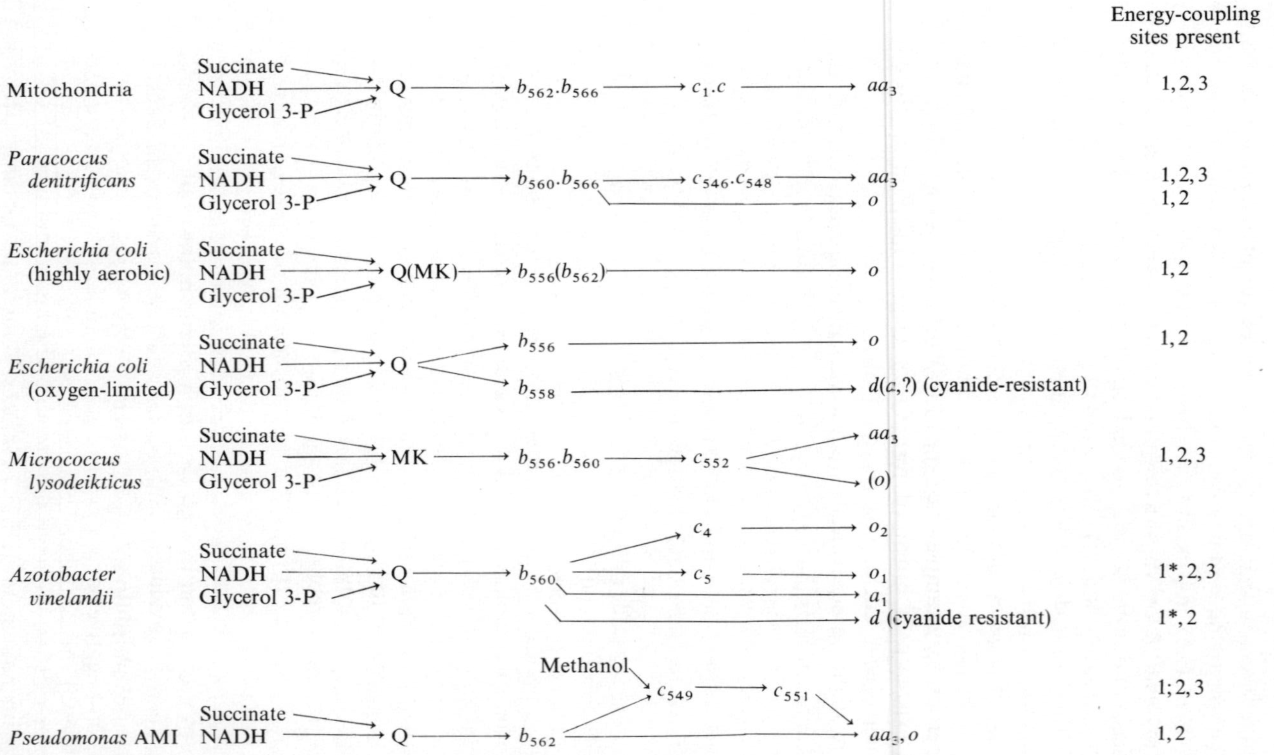

Mitochondria
Succinate, NADH, Glycerol 3-P → Q → $b_{562}.b_{566}$ → $c_1.c$ → aa_3 — 1, 2, 3

Paracoccus denitrificans
Succinate, NADH, Glycerol 3-P → Q → $b_{560}.b_{566}$ → $c_{546}.c_{548}$ → aa_3 / → o — 1, 2, 3 ; 1, 2

Escherichia coli (highly aerobic)
Succinate, NADH, Glycerol 3-P → Q(MK) → $b_{556}(b_{562})$ → o — 1, 2

Escherichia coli (oxygen-limited)
Succinate, NADH, Glycerol 3-P → Q → b_{556}, b_{558} → o, $d(a,?)$ (cyanide-resistant) — 1, 2

Micrococcus lysodeikticus
Succinate, NADH, Glycerol 3-P → MK → $b_{556}.b_{560}$ → c_{552} → aa_3, (o) — 1, 2, 3

Azotobacter vinelandii
Succinate, NADH, Glycerol 3-P → Q → b_{560} → c_4, c_5 → o_2, o_1, a_1, d (cyanide resistant) — 1*, 2, 3 ; 1*, 2
Methanol → c_{549} → c_{551}

Pseudomonas AM1
Succinate, NADH → Q → b_{562} → c_{549} → c_{551} → aa_3, o — 1; 2, 3 ; 1, 2

Energy-coupling sites present (right-hand column)

*Under highly aerobic conditions site 1 is absent from nitrogen-fixing Azotobacter vinelandii.

Figure 7.2. Mitochondrial and heterotrophic bacterial respiratory chains. Cytochromes are designated by their letter and, wherever possible in the case of b- and c-types, a subscript recording their wavelength maxima in low temperature reduced minus oxidized difference spectra. Parentheses indicate redox carriers of low concentration or activity. Based on data in Jones (1977, 1982), and Jurtshuk and Yang (1980).

positive bacteria and that MK is also found in some Gram-negatives. The most marked differences, though, are found in the cytochrome oxidase system where the mitochondrial cytochrome aa_3 may be replaced by up to three bacterial cytochrome oxidases, namely aa_3, o and d, all of which bind carbon monoxide. Other potential oxidases are cytochromes a_1 and $c_{(CO)}$.

(2) The addition or deletion of one or more redox carriers. For example cytochrome c and either energy-dependent or energy-independent transhydrogenases may be absent.

(3) The substitution of one carrier by another with markedly different properties, for example one type of transhydrogenase by another.

(4) The existence of branched respiratory chains. Bacteria all display considerable branching at the level of the primary dehydrogenases where reducing equivalents from a variety of substrates other than NADH are channelled into the respiratory chain, and their terminal pathways (cytochrome oxidase) may be either linear or branched. If oxygen is replaced as the terminal electron acceptor by fumarate or nitrate under anaerobic conditions, branching always occurs. Examples of branched respiratory chains are depicted in Figure 7.2. Of particular interest are the dual terminal branches, one of which is usually more resistant to cyanide inhibition than the other, e.g. *Azotobacter vinelandii* and *Escherichia coli* grown under oxygen-limited conditions display this pattern. It seems to be generally the case that in these complex systems one branch comprises cytochrome c and cytochrome oxidases aa_3 or o, whereas the other branch has a b-type cytochrome and terminates with cytochrome oxidase aa_3, d or o.

The composition of the respiratory chain of an individual bacterial species is susceptible to growth conditions, especially to the availability of molecular oxygen. Growth under oxygen limitation generally leads to the partial or complete replacement of cytochrome oxidase aa_3 by o (e.g. *P. denitrificans* and *A. eutrophus*) or to increased synthesis of cytochrome oxidase d relative to o and/or $c_{(CO)}$ (e.g. *E. coli*—see Fig. 7.2, *Klebsiella pneumoniae* and *Haemophilus parainfluenzae*). These organisms thus combat oxygen deprivation by synthesizing higher concentrations of alternative cytochrome oxidases which possibly have greater affinities for molecular oxygen and/or higher turnover numbers. As previously noted, many Gram-negative bacteria contain both ubiquinone and menaquinone.

7.2.1 Spatial arrangement of chain components

The spatial arrangement of components of the respiratory chain within the plasma membrane has been investigated using whole cells, spheroplasts, protoplasts and by comparing right-side-out membrane vesicles with inside-out (everted) vesicles (i.e. the cytoplasmic side of their membrane is on the outside of the vesicle). A combination of techniques, including electron microscopy, accessibility of redox components to non-penetrating substrates and effectors, to antibodies and to proteolytic enzymes, has been applied and valuable information secured, although interpretation of results requires caution. Evidence suggests that the active sites of NADH dehydrogenase and transhydrogenase are located on the cytoplasmic side of the membrane, as are those of a number of other dehydrogenases and cytochrome oxidase. In contrast, cytochrome c and methanol dehydrogenase are oriented on the outer, periplasmic surface. The lipid-soluble quinones are believed to be deeply embedded in the membrane, while the location of the iron-sulphur proteins and other cytochromes remains to be ascertained. These findings are consistent with the concept of a transmembrane respiratory chain as envisioned in the chemiosmotic hypothesis.

7.2.2 Control of bacterial respiration

Both coarse and fine controls operate on bacterial respiration, the former by the repression and induction of redox carrier synthesis, while the latter is effected by the intrinsic kinetic properties of the respiratory chain and its associated energy conservation mechanism. The concentrations of quinones and cytochromes in the coupling membrane appear to be in excess, with the rate-limiting step of the chain residing in the primary dehydrogenations that feed reducing equivalents into the sequence. The dehydrogenase enzymes themselves are subject to various types of control, among which have been noted modulation by activators and inhibitors as well as by substrate availability. Further, the rate of the cytochrome-oxidase-catalysed reaction is influenced by the prevailing concentration of molecular oxygen.

Appropriate bacterial cell suspensions and membrane vesicle preparations also display the classical respiratory control characteristic of yeast and mammalian mitochondria, i.e. their respiration is stimulated by the addition of ADP plus P_i or of uncoupling agents or ionophorous antibiotics which collapse the protonmotive force (Δp).

The presence of branched respiratory pathways undoubtedly confers some flexibility on the precise route of electron transfer available to an organism and affords scope for circumventing the effects of potentially harmful environments and for securing maximum benefit from others.

Typical of branched respiratory chains are the different stoichiometries of ATP production characteristic of the two terminal pathways; the one associated with the lowest ATP formation permits the highest rate of electron transfer. This property is of particular interest in relation to *Azotobacter vinelandii* and other azotobacters which need to protect their oxygen-sensitive nitrogenase system. In the process of 'respiratory protection' a high environmental oxygen concentration is scavenged by a rapid respiratory rate achieved via cytochrome oxidase d, yet electron transfer can be switched to the more efficiently coupled pathway via cytochromes c_4 and c_5 when maximum energy conservation is paramount.

7.3 Proton-translocating and energy-transducing sites

Experiments with mitochondria and with suspensions of intact cells or protoplasts, spheroplasts and membrane vesicles prepared from a variety of bacteria, have demonstrated that pulses of dissolved oxygen cause proton extrusion, except in the case of inside-out vesicles, when the protons move inwards. Under appropriate experimental conditions it is possible to measure the $\rightarrow H^+/O$ ($\rightarrow H^+/2e^-$) quotients for such preparations, i.e. the stoichiometric ratio of protons extruded (or taken up by inside-out vesicles) per atom of oxygen utilized. With intact cells oxidizing endogenous metabolites $\rightarrow H^+/O$ quotients of approximately 4, 6 and 8 have been obtained depending upon the composition of the respiratory chain involved. For example, *Escherichia coli* and other organisms that lack cytochrome c and are devoid of transhydrogenase activity display maximum $\rightarrow H^+/O$ quotients of about 4. Mitochondria oxidizing exogenous substrates via internal $NADP^+$, NAD^+ and flavin yield $\rightarrow H^+/O$ quotients of about 8, 6 and 4 respectively. Certain bacteria under anaerobic conditions display $\rightarrow H^+/O$ quotients of around 2 and 4 for the oxidation of NADH by fumarate and nitrate respectively.

These results led to the conclusion that in the process of electron transfer each coupling site expels $2H^+$. A mechanism which could account for the observations is the transmembrane redox loop (Figure 7.3) comprising a hydrogen carrier (flavin, quinone) followed by an electron carrier (Fe–S protein, cytochrome). An internally-sited NAD^+-linked dehydrogenase feeds reducing equivalents (2H) to the hydrogen carrier which transfers

D

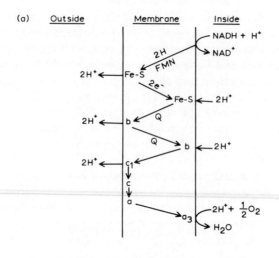

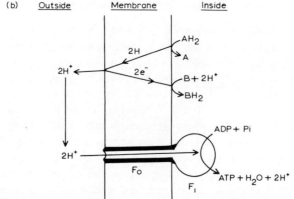

Figure 7.3 (*a*) Proton-translocating redox loops of the respiratory chain showing the arrangement of carriers as originally proposed by Peter Mitchell. (*b*) Diagrammatic general case representation of proton-translocating redox loop and proton-translocating ATP phosphohydrolase (ATP synthetase) as envisaged by the chemiosmotic hypothesis.

them outwards, $2H^+$ are extruded to the exterior and $2e^-$ transferred inwards. The redox components of sites 1 and 2 are compatible with a redox loop mechanism whereas site 3 (cytochrome oxidase), which possesses only an electron carrier function, is not. To overcome this problem Mitchell (1975) proposed that sites 2 and 3 are combined to furnish a protonmotive quinone cycle which, in a complicated sequence involving quinone, free-radical semiquinone and quinol forms of either ubiquinone or menaquinone, effects the extrusion of $4H^+$ as two pairs of protons in the course of two successive single-electron transfers from an Fe–S centre in the NADH-ubiquinone reductase to cytochrome c (Figure 7.4). Cytochrome oxidase then serves as a vectorial electron carrier from cytochrome c, to catalyse reaction with molecular oxygen at the interior surface of the membrane.

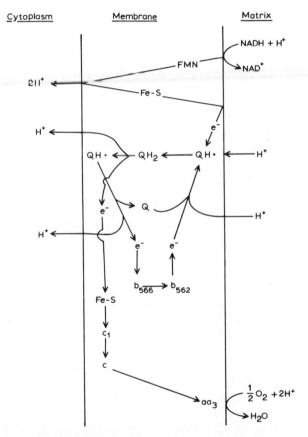

Figure 7.4. The mitochondrial protonmotive quinone cycle which has been proposed to operate between NADH dehydrogenase and the cytochrome system. The quinone Q (which may be ubiquinone, plastoquinone or other analogous compound) undergoes reduction via the semiquinone AH to hydroquinone QH_2 which is then re-oxidized. There are two such cycles per NADH and thus the overall $\rightarrow H^+/O$ quotient is 6. Experimental verification of the protonmotive quinone cycle remains to be secured. (Note that the transfer of single electrons is implied by the relevant arrows shown within the membrane and that each electron carrier undergoes a reduction-oxidation cycle in accepting and donating an electron.)

This ingenious proposal, which accounts for the observed phenomena, awaits experimental verification.

The problem of the number of proton-translocating sites and their location within the respiratory chain has been tackled by measuring $\rightarrow H^+/O$ quotients after adding substrates (physiological or artificial) which react with different redox components of the chain, often combined with the judicious use of selected inhibitors to prevent activity in specific regions of the chain. By this means the $\rightarrow H^+/O$ quotients associated with isolated segments of the electron transfer system have been determined. The results indicate that mitochondria possess three sites, at the levels of NADH dehydrogenase (site 1), the quinone-cytochrome c system (site 2) and cytochrome oxidase (site 3). A fourth site (site 0) exists at the level of transhydrogenase but, as already noted (p. 85), is believed to function principally in reversed electron transfer and not to play a significant role in oxidative phosphorylation. In bacteria the presence of sites 0 and 3 depends on the capacity of the organism to synthesize respectively a transhydrogenase and a cytochrome c of high redox potential linked to cytochrome oxidase aa_3 or o (Figure 7.2).

While the energy-coupling sites of mitochondria can be ascertained by measuring ATP/O (ATP/2e$^-$) quotients by techniques similar to those just discussed for $\rightarrow H^+/O$ quotients, but adding ADP plus P_i and determining the ATP formed, it is not possible to do likewise with intact bacteria because they lack an adenine nucleotide translocase. Consequently, exogenous ADP cannot be phosphorylated nor can ATP be hydrolysed. For this reason, inside-out membrane vesicles have generally been used to measure bacterial ATP/O quotients, although technical problems usually result in low efficiencies of coupling with ATP/O quotients for NADH oxidation rarely exceeding 1.5. Evaluation of sites has therefore been carried out by identifying those regions of the carrier chain which (1) display stimulated respiration when uncoupling agents are added, (2) can effect ATP synthesis or ion transport at the expense of forward electron transfer, and (3) show reversed electron transfer driven by ATP hydrolysis or by forward electron transfer through a different energy transduction site. Results obtained by these methods using mitochondria and bacteria correlate well with those secured by $\rightarrow H^+/O$ measurements. They can also be supplemented by *in-vivo* measurements of maximum molar growth yields (Chapter 4) with energy-limited continuous cultures. Micro-organisms which possess a terminal electron transfer pathway comprising cytochrome c and cytochrome oxidase aa_3 or o display molar growth yields that are about 50% greater than organisms which lack cytochrome c, corresponding to the presence of sites 1, 2 and 3 in the former bacteria but only sites 1 and 2 in the latter.

A few final words of caution are apposite concerning mechanisms of oxidative phosphorylation. Fairly recent work on the redox loops and proton translocation in mitochondria has cast doubt on the values of $\rightarrow H^+/O$ quotients obtained in oxygen-pulse experiments. When alternative methods of measurement, involving the addition of an oxidizable substrate to mitochondrial suspensions, are adopted, $\rightarrow H/O$ quotients as high as 12, 8 and 4 have been secured for the oxidation of NAD^+-linked substrates, succinate and ferrocyanide respectively. Such high values cannot be accommodated in the simple redox loop mechanism of proton translocation discussed earlier in this chapter. This field of research has generated a considerable literature, usually characterized by a lack of unanimity of interpretation, and to keep abreast of developments the general reader will be served best by a judicious choice of the latest review articles.

CHAPTER EIGHT

ENERGY CONSERVATION IN CHEMOHETERO-TROPHIC ANAEROBIC METABOLISM

Obligate anaerobes and facultative organisms under anaerobic conditions gain ATP by substrate-level and/or electron transport (transfer) phosphorylation. In the present chapter the reactions leading to ATP formation will be considered.

8.1 Energy conservation via substrate-level phosphorylation

The majority of anaerobes secure their energy via substrate-level phosphorylation (Gottschalk and Andreesen, 1979). Despite the variety of carbon substrates fermented, the number of substrate-level phosphorylation reactions identified is small. Thus the principal 'high-energy' intermediates are few, e.g. acyl-SCoA, acetyl phosphate, 1, 3-bisphosphoglycerate, phosphoenolpyruvate, carbamoyl phosphate, and formyltetrahydrofolate. With the exception of acyl-SCoA and formyltetrahydrofolate all contain high-energy phosphate functions which can be used to phosphorylate ADP directly. Possibly the most important of these intermediates is acetyl-SCoA which, under anaerobic conditions, cannot be oxidized via the tricarboxylic acid cycle but yields energy by substrate-level instead of electron transport phosphorylation.

In anaerobiosis energy conservation from acetyl-SCoA in prokaryotes occurs via reactions catalysed by the enzymes *phosphotransacetylase* and *acetate kinase*:

$$CH_3CO.SCoA + P_i \rightleftharpoons CH_3CO.OPO_3H_2 + CoASH$$
$$\Delta G^{0'} = +9.0 \, kJ \, mol^{-1}$$
$$CH_3CO.OPO_3H_2 + ADP \rightleftharpoons CH_3COOH + ATP$$
$$\Delta G^{0'} = -13.0 \, kJ \, mol^{-1}$$

This pair of enzymes is found in all anaerobic bacteria that produce acetyl-SCoA in their metabolism and employ it for ATP synthesis. A few aerobic bacteria, e.g. *Azotobacter vinelandii* and *Acetobacter xylinum,* also possess them. Regulatory roles have been shown for the phosphotrans-acetylase of *Escherichia coli* and the acetate kinase of *Veillonella alcalescens.* These enzymes do not seem to mediate in eukaryotic metabolism but the formation of acetate from acetyl-SCoA in anaerobic protozoa is catalysed by an *acetate thiokinase:*

$$CH_3CO.SCoA + ADP + P_i \rightleftharpoons CH_3COOH + CoASH + ATP$$
$$\Delta G^{0'} = -4.0\,kJ\,mol^{-1}$$

which also functions with GDP. The energy of the thioester bond is thus conserved in either ATP or GTP.

8.1.1 Anaerobic formation of acetyl-SCoA

Under anaerobiosis the NADH-dependent pyruvate dehydrogenase multi-enzyme complex characteristic of aerobic metabolism (p. 39) does not operate (except in *Rhodospirillum rubrum*) and acetyl-SCoA is formed from pyruvate via two other reactions. It can also arise from sources other than pyruvate. Some anaerobes, such as clostridia, sulphate-reducing bacteria, peptococci, sarcina, and hydrogen-forming protozoa, possess a *pyruvate: ferredoxin oxidoreductase* which catalyses the reaction

$$CH_3CO.COOH + CoASH + 2Fd_{ox} \rightleftharpoons CH_3CO.SCoA + CO_2 + 2Fd_{red}$$
$$\Delta G^{0'} = -19.2\,kJ\,mol^{-1}$$

Ferredoxin (Fd) is a one-electron carrier with an extremely negative redox potential ($E_0' = -400\,mV$) virtually equivalent to the standard hydrogen electrode. The reduced ferredoxin can then be used in other reactions. In cyanobacteria the pyruvate: ferredoxin oxidoreductase seems principally to have an anabolic function, producing reduced ferredoxin for the reduction of NADP or nitrogen, while in *Clostridium kluyveri* it operates to form pyruvate from acetyl-SCoA and CO_2 rather than in the reverse direction.

The second enzyme which yields acetyl-SCoA from pyruvate is *pyruvate: formate lyase* which catalyses the thiolysis of the keto acid and is characteristic of anaerobically-grown Enterobacteriaceae and anaerobic fermenting Rhodospirillaceae. The enzyme is active only under anaerobic conditions.

$$CH_3CO.COOH + CoASH \rightleftharpoons CH_3CO.SCoA + HCOOH$$
$$\Delta G^{0'} = -16.3\,kJ\,mol^{-1}$$

Some clostridia possess the lyase but employ it principally to secure formate for the synthesis of one-carbon units rather than to obtain acetyl-SCoA for ATP generation.

Many micro-organisms have a coenzyme A-dependent *aldehyde dehydrogenase (acylating)* which effects the reversible dehydrogenation of acetaldehyde to acetyl-SCoA.

$$CH_3CHO + CoASH + NAD^+ \rightleftharpoons CH_3CO.SCoA + NADH + H^+$$
$$\Delta G^{0\prime} = -17.5\,kJ\,mol^{-1}$$

In most cases the enzyme functions in the formation of ethanol, or in an analogous reaction with butyryl-SCoA leading to butanol production; under these circumstances the high-energy function of the thioester is lost. However, in a few instances, e.g. *Clostridium kluyveri* in the fermentation of ethanol, the enzyme functions to produce acetyl-SCoA and utilizes either NAD^+ or $NADP^+$ as electron acceptor.

Important in fatty acid and poly-β-hydroxybutyrate metabolism are β-ketothiolases which permit the formation of acetyl-SCoA from the CoA-esters of 2-oxoacids. The equilibrium of the reaction greatly favours thiolysis. For example, the reaction catalysed with acetoacetyl-SCoA

$$CH_3CO.CH_2CO.SCoA + CoASH \rightleftharpoons 2CH_3CO.SCoA$$

has $\Delta G^{0\prime} = -25.1\,kJ\,mol^{-1}$. However, under anaerobiosis the main role of the enzyme is for the synthesis of acetoacetyl-SCoA from acetyl-SCoA in the formation of butyric acid in the butyric acid fermentation. There are a few instances of the reverse reaction in anaerobes, for example when *Clostridium kluyveri* grows on crotonate β-ketothiolase operates to yield acetyl-SCoA for ATP synthesis via phosphotransacetylase and acetate kinase.

Anaerobic metabolism of amino acids such as threonine and leucine can yield acyl-SCoA derivatives other than acetyl-SCoA, e.g. propionyl-SCoA and butyryl-SCoA, and it seems likely that these compounds are also able to yield ATP via analogous reactions of a phosphotransacetylase and acyl kinase.

8.1.2 Acetyl phosphate formation

We have already seen that acetyl-SCoA is the major precursor of acetyl phosphate via the action of phosphotransacetylase. However, a few

anaerobes can produce acetyl-SCoA from other compounds. Thus
Lactobacillus delbruckii possesses a flavin adenine dinucleotide (FAD)-
linked pyruvate dehydrogenase which yields acetyl phosphate directly

$$CH_3CO.COOH + FAD + P_i \rightleftharpoons CH_3CO.OPO_3H_2 + FADH_2 + CO_2$$

Furthermore, the heterolactic acid fermenting bacteria possess phosphoketolase (p. 37) which cleaves xylulose 5-phosphate to yield acetyl
phosphate and glyceraldehyde phosphate; when pentose is the fermentation substrate acetyl phosphate is used to generate ATP, whereas when
hexose is fermented the acetyl phosphate is reduced to ethanol and the high-
energy function is lost (p. 39).

The *Bifidobacteria* present an interesting example of organisms which
employ phosphoketolase to cleave both xylulose 5-phosphate and fructose
6-phosphate, yielding acetyl phosphate and glyceraldehyde 3-phosphate or
erythrose 4-phosphate respectively. Starting with 2 moles of glucose, after
phosphorylation using 2 ATP, the two hexose phosphates are converted, by
the combined action of phosphoketolase, transaldolase and transketolase,
to 2 moles of glyceraldehyde 3-phosphate and 3 moles of acetyl phosphate.
By reactions common to glycolysis the 2 triose phosphates yield 2 lactates
in a balanced oxido-reduction and 4 ATP by substrate-level phosphorylation. With 3 ATP derived from 3 acetyl phosphate the net energy yield is
2.5 ATP per mole of glucose fermented (Figure 8.1).

$$2 \text{ glucose} + 5 \text{ ADP} + 5P_i \rightarrow 2 \text{ lactate} + 3 \text{ acetate} + 5 \text{ ATP}$$

The ATP yield is thus higher than that of glycolysis or the heterolactic
fermentation.

8.1.3 1,3-Bisphosphoglycerate and phosphoenolpyruvate formation

The reactions leading from glyceraldehyde 3-phosphate to pyruvate are
common to the four principal pathways of glucose metabolism (Chapter
3). As previously described, these reactions embrace two substrate-level
phosphorylation reactions yielding 2 moles of ATP per mole of triose
phosphate metabolized (see pp. 23–27). The oxidation of glyceraldehyde
3-phosphate to 1,3-bisphosphoglyceraldehyde generates NADH which
anaerobically is used to reduce pyruvate, acetyl-SCoA or, in the case of
saccharolytic clostridia, protons, hydrogen being released via NADH:
ferredoxin reductase and ferredoxin hydrogenase.

8.1.4 Carbamoyl phosphate formation

Micro-organisms that are able to grow on arginine as the energy source are few. They produce derivatives of urea (ureido compounds of general formula $RNH.CO.NH_2$) as intermediates and these compounds are further dissimilated by phosphorolysis to yield carbamoyl phosphate and the respective amide or amine. The carbamoyl phosphate is then reacted with ADP under the influence of *carbamate kinase* (*ATP-carbamate phosphotransferase*) to yield ATP and carbamate. The overall reactions are:

$$
\begin{array}{ccc}
NH_2 & NH_2 & NH_2 \\
| & | & | \\
C{=}NH & C{-}O & H_2O_3P \quad O \quad C{-}O \\
| \quad\quad H_2O & | \quad\quad P_i & + \\
NH \quad \longrightarrow & NH \quad \longrightarrow & NH_2 \\
| \quad\quad NH_3 & | & | \\
(CH_2)_3 & (CH_2)_3 & (CH_2)_3 \\
| & | & | \\
HC{-}NH_2 & HC{-}NH_2 & HC{-}NH_2 \\
| & | & | \\
COOH & COOH & COOH
\end{array}
$$

L-arginine   L-citrulline   carbamoyl phosphate
+
L-ornithine

$$NH_2$$
$$|$$
$$H_2O_3P{-}O{-}C{=}O + ADP \rightleftharpoons NH_3 + CO_2 + ATP$$

Sum: $arginine + ADP + P_i + H_2O \rightarrow ornithine + CO_2 + 2NH_3 + ATP$

$$\Delta G^{0'} \simeq -20\,kJ\,mol^{-1}$$

The first two reactions of the sequence are catalysed by *arginine deimidase* (*iminohydrolase*) and *ornithine carbamoyl transferase* respectively.

Streptococcus faecalis accomplishes these reactions and it has been shown that the growth yield on arginine corresponds to 1 ATP per mole of substrate fermented. The anaerobic degradation of allantoin by *Streptococcus allantoicus* and enterobacteria also proceeds via carbamoyl phosphate enabling ATP synthesis to occur.

8.1.5 Formyltetrahydrofolate formation

Reaction of the compound N^{10}-formyltetrahydrofolate (N^{10}-formyl FH_4) with ADP and P_i is effected by the enzyme *formyltetrahydrofolate synthetase* yielding formate, tetrahydrofolate and ATP:

$$formyl\ FH_4 + ADP + P_i \rightleftharpoons formate + FH_4 + ATP$$

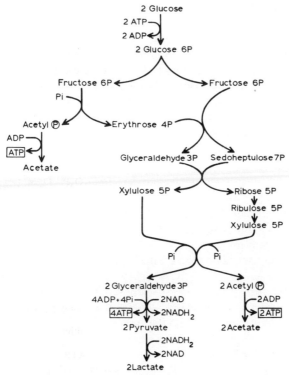

Figure 8.1. The *Bifidobacterium bifidum* fermentation of glucose which displays a net yield of 2.5 ATP per mole of glucose fermented.

Although this enzyme is of fairly widespread occurrence its major role seems to be in the synthesis of formyl FH_4 from formate and tetrahydro-folate rather than in ATP formation. Indeed *Clostridium cylindrosporum*, which produces formyl FH_4 as an intermediate in purine fermentation, seems to be the only well-documented example of ATP generation by this route.

8.2 Energy conservation via electron transport phosphorylation

Formerly, ATP synthesis via electron transport phosphorylation was believed to be associated only with aerobic respiratory processes, and hence referred to as oxidative phosphorylation (Chapter 7). However, in more recent times it has been shown conclusively that electron transfer coupled to ATP synthesis also occurs in some facultative and obligate anaerobes. Many obligate anaerobes can use hydrogen as an electron donor in mixed fermentations with either fumarate, nitrate, sulphate or CO_2 (HCO_3^-) as electron acceptor and derive energy for growth from the process. Because substrate-level phosphorylation involving H_2 is unknown, it was inferred that electron transport phosphorylation must be involved. Direct experimental evidence has now been obtained for a number of these systems and here we shall consider some representative examples.

8.2.1 Fumarate reduction: fumarate reductase

An enzyme which reduces fumarate to succinate and is genetically distinct from succinate dehydrogenase is present in many facultative and obligate anaerobes. A prime example is the anaerobic *Vibrio succinogenes* which grows with H_2 as the energy source using fumarate as the electron acceptor, but other organisms include *Desulfovibrio gigas*, *Clostridium formicoacetium*, Propionibacteria, some Bacteroides and the facultative *Escherichia coli* and *Proteus rettgeri* (Kroger, 1977, 1978). A few eukaryotes also derive energy from the reduction of fumarate to succinate. Thus some anaerobic protozoa and parasitic helminths (*Ascaris lumbricoides, Fasciola hepatica*) employ fumarate as the terminal electron acceptor of the mitochondrial respiratory chain in place of oxygen.

Under anaerobiosis fumarate is formed from a variety of sources, such as malate, aspartate, pyruvate + CO_2, and is therefore available as an electron acceptor. The electron donors hydrogen and formate are formed as fermentation products and NADH is produced in metabolism. It must be borne in mind however that the reduction of fumarate to succinate in many anaerobes fulfils an anabolic purpose, providing the organism with succinate for tetrapyrrole synthesis, rather than serving in an energy-generating capacity.

Fumarate reductase is a membrane-bound iron-sulphur flavoprotein enzyme with a covalently-linked FAD prosthetic group which consists of two subunits ($M = 79k$ and $31k$ respectively). It is not constitutive in most bacteria, and anaerobic growth on fumarate or fumarate precursors induces the enzyme. In the membrane (the mitochondrial membrane of eukaryotes) fumarate reductase is associated with electron carriers that link it to electron donors such as H_2, formate, lactate, glycerol phosphate or NADH. Although some variation is encountered in the carrier systems of different organisms and the electron donors they employ, the lipophilic compound menaquinone (or 2-demethylmenaquinone) and a *b*-type cytochrome are characteristic electron carriers of the bacterial fumarate reductase system. The mitochondria of the anaerobic helminths which reduce fumarate contain rhodoquinone rather than ubiquinone.

Menaquinone ($E'_0 = -74$ mV) has a standard redox potential more negative than the succinate/fumarate couple ($E'_0 = +33$ mV) whereas ubiquinone ($E'_0 = +133$ mV) is considerably more positive, according with the observation that ubiquinone operates in the oxidation of succinate whereas menaquinone functions in fumarate reduction. When NADH is the electron donor the redox span of the NADH/NAD$^+$ ($E'_0 = -320$ mV) and succinate/fumarate couples, namely 353 mV, is sufficient to permit the synthesis of 1 ATP from ADP and P_i by electron transport phosphorylation. Likewise the couples H$_2$/2H$^+$ ($E'_0 = -414$ mV) and formate/CO$_2$($E'_0 = -432$ mV) ensure an adequate redox span with the succinate/fumarate couple for ATP synthesis.

Type b cytochromes are generally but not universally associated with fumarate reductase activity. Thus examples of fumarate reduction are known where the presence of cytochrome b is mandatory (*Bacteroides fragilis*) and where it is not (a cytochromeless mutant of *Escherichia coli*). It is significant however that the cytochromeless mutant of *E. coli* reduces fumarate at a lower rate and grows more slowly on glycerol-fumarate medium than the wild-type, suggesting the cytochrome b is required for high rates of fumarate reduction. *Vibrio succinogenes* has two b-type cytochromes associated with fumarate reduction by formate, one linked to formate dehydrogenase and the other to the fumarate reductase (Figure 8.2).

Experimental evidence for ATP synthesis in the course of fumarate reduction stems from three types of observation. First, particulate preparations of *Desulfovibrio gigas* couple the phosphorylation of ADP with the reduction of fumarate by H$_2$. Similar findings have been made with spheroplasts of *Vibrio succinogenes*, while inside-out vesicles prepared from *E. coli* effect fumarate-dependent proton translocations. Second, organisms growing anaerobically on substrates such as H$_2$ plus fumarate, with succinate as the sole end product, are unable to secure ATP by substrate-level phosphorylation and must therefore employ electron transport phosphory-

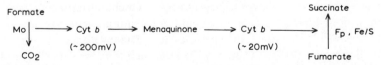

Figure 8.2   Electron transport from formate to fumarate in *Vibrio succinogenes*. After Kröger (1978). Abbreviations: Mo, molybdoprotein; Fp, Fe/S, FAD–iron–sulphur protein.

lation. Third, the growth yields of organisms fermenting carbohydrates to acetate, propionate and succinate (e.g. *Selenomonas ruminantium, Propionibacterium freudenreichii* and *Bacillus fragilis*) are much higher than those of homolactic and ethanol-producing bacteria, e.g. 50–65 compared with 20–35 g dry weight per mole of glucose, indicating that fumarate reduction is linked with ATP formation.

Experiments with permeant and impermeant electron donors provided evidence that the fumarate reductase of *Vibrio succinogenes* is located on the inner side of the cytoplasmic membrane, while the formate dehydrogenase associated with fumarate reduction is on the outer side, denoting that fumarate reduction is a transmembrane redox process. During the reduction of 1 molecule of fumarate with formate two protons are released on the outer side and two protons are consumed on the inner side of the membrane.

8.2.2 Nitrate reduction: nitrate reductase

The ability to reduce nitrate to nitrite

$$NO_3^- + H_2 \rightarrow NO_2^- + H_2O \quad \Delta G^{0'} = -163 \, \text{kJ} \, \text{mol}^{-1}$$

is widely encountered in facultatively anaerobic bacteria (e.g. *Escherichia coli, Klebsiella pneumoniae, Paracoccus denitrificans* and various denitrifying species of *Pseudomonas*) and also in a few obligate anaerobes (e.g. *Veillonella alcalescens, Selenomonas ruminantium* and *Clostridium perfringens*). In most of these organisms the nitrite formed is not further reduced and is exported from the cells.

The nitrate reduction system comprises a membrane-bound nitrate reductase together with associated dehydrogenases and electron carriers. Several reductases have been isolated and investigated. That from *Escherichia coli* is a molybdenum-containing iron-sulphur protein composed of two different subunits, $\alpha(M = 155\text{k})$ and $\beta(M = 63\text{k})$. The carrier is a b-type cytochrome ($M = 19\text{k}$) specific for nitrate reduction and designated the γ subunit (cyt $b_{556}^{NO_3^-}$). The subunit stoichiometry of the reductase complex is $\alpha\beta\gamma$ or $\alpha\beta\gamma_2$ and it is believed to exist as a tetramer of this in the cytoplasmic membrane with the α and β subunits located on the inside and the cytochrome $b_{556}^{NO_3^-}$ on the outer (periplasmic) surface, the complex thereby spanning the membrane. There are 12 atoms each of Fe and labile S per Mo atom in the complex and both the Fe–S and Mo centres are involved in respiratory activity.

The α-subunit catalyses the reduction of nitrate and the consumption of

$2H^+$ on the inner (cytoplasmic) side of the membrane, while the anaerobic oxidation of electron and hydrogen donors occurs at the outer surface with appropriate proton release. Haem-deficient mutants do not display these redox reactions whereas mutants lacking both ubiquinone and mena-quinone do, indicating that cytochrome but not quinone is obligatory for nitrate reduction. The evidence supports the concept of proton trans-location via a redox arm system with nitrate reductase catalysing trans-membrane electron flow.

The electron donors for nitrate reduction include NADH, succinate, lactate, formate, glycerol phosphate and hydrogen and the relevant membrane-bound dehydrogenases for these substrates are present under the appropriate growth conditions. The most detailed investigations have been carried out with the formate dehydrogenase of *Escherichia coli*, which is a selenium-molybdenum-haem protein comprising three subunits (α, β and γ with M 110k, 32k and 20k respectively) together with a substantial excess of iron–sulphur protein. The Se, Mo and haem are present in equimolar amounts with the Mo possibly functioning as the first redox centre in the sequence. As with nitrate reductase, it seems that the α and β subunits span and membrane and possibly formate oxidation occurs on the inner side. However, unlike the coliform enzyme, the formate dehydro-genase of *Vibrio succinogenes* lacks both selenium and iron-sulphur proteins and is oriented on the outer side of the cytoplasmic membrane.

Evidence for electron transport phosphorylation in anaerobic nitrate reduction is derived from two sources. First, *Veillonella alcalescens* grows anaerobically on H_2 and nitrate as the sole energy source yielding only nitrite as product, so that energy must be secured from this conversion. Molar growth yields determined on various bacteria with nitrate as the terminal electron acceptor support the belief that nitrate reduction yields ATP. Second, measurements of $\rightarrow H^+/NO_3^-$ ratios for the oxidation of NAD-linked substrates or formate by different organisms suggest there are two proton translocating segments, each associated with $2H^+$, giving an overall $\rightarrow H^+/NO_3^-$ of $\simeq 4$ and a Δp adequate to drive ATP synthesis. Determinations of P_i esterification during nitrate reduction to nitrite by sub-cellular preparations of *Escherichia coli* and *Micrococcus denitrificans* confirm that ATP synthesis occurs.

8.2.3 Nitrite reduction

There are two pathways for the reduction of nitrite produced in metabolism by the action of nitrate reductase, leading respectively to the formation of ammonia and molecular nitrogen. The first type is 'assimilatory', characteristic of micro-organisms that utilize nitrate as a nitrogen source for growth, while the second is 'respiratory' and found in those organisms which employ nitrate as an electron acceptor alternative to oxygen. As the nitrite ion is cytotoxic, the physiological role of nitrite reduction may also embrace detoxication of the environment.

Assimilatory nitrite reductase. This enzyme is soluble, NAD^+-dependent and found in facultative anaerobes such as *E. coli* and *Klebsiella pneumoniae*. It permits them to secure ammonia from nitrite without the necessity for nitrogen fixation (i.e. it eliminates the nitrogen cycle sequence $NO_2^- \rightarrow NO \rightarrow N_2O \rightarrow N_2$) as follows:

$$NO_2^- + 3NADH + 4H^+ \rightarrow NH_3 + 3NAD^+ + 2H_2O$$

The reaction thus effects the reoxidation of NADH and thereby alters the balance of the fermentation in favour of acetate rather than ethanol formation. Energy is thereby conserved via acetyl phosphate even though the nitrite reductase reaction does not itself conserve energy.

The reduction of nitrite to ammonia, coupled with the oxidation of hydrogen, in the sulphate-reducing bacterium *Desulfovibrio gigas* results in a transmembrane proton gradient of sufficient magnitude to drive ATP synthesis via a proton-translocatory ATPase. The observed $\rightarrow H^+/2e^-$ quotient was 2.0 although the $P/2e^-$ ratio was low (about 0.4), possibly reflecting the inefficient functioning of the proton pump coupled to electron transfer in the cell-free system used.

Respiratory nitrite reductase. Denitrifying organisms, that is those utilizing nitrate as an electron acceptor in place of oxygen and producing molecular nitrogen in the process, export the nitrite formed by the action of nitrate reductase. Outside the cytoplasmic membrane the nitrite is reduced by the action of two periplasmic enzymes, *nitrite reductase*, which produces nitrous oxide as intermediate, and *nitrous oxide reductase* which yields dinitrogen, thus

$$2NO_2^- + 6H^+ + 4e^- \rightarrow N_2O + 3H_2O$$
$$N_2O + 2H^+ + 2e^- \rightarrow N_2 + H_2O$$

The nitrite reductase is a cytochrome cd_1 ($M = 122k$) composed of two identical subunits each of which carries one molecule of haems c and d_1. The enzyme receives electrons from the Q-cytochrome b segment of the respiratory chain via two membrane-bound c-type cytochromes, a periplasmic blue copper protein *azurin* and a periplasmic cytochrome c_{551}. As will be evident from the above equations, the formation of one molecule of nitrogen from $2NO_2^-$ involves the consumption of $8H^+$ on the periplasmic side of the membrane and the transfer of $6e^-$. Experimentally it has been shown that anaerobic respiration of nitrite by *Paracoccus denitrificans* is associated with net proton translocation and ATP synthesis. However, at concentrations of 7 μM and higher, nitrite acts as an uncoupler, rendering the *P. denitrificans* membrane permeable to protons and collapsing the proton gradient.

Molar growth yield studies showed that the reactions catalysed by nitrite reductase and nitrous oxide reductase conserve equivalent amounts of energy per electron transferred.

8.2.4 Sulphate-reducing bacteria

The ability to utilize sulphate as a terminal electron acceptor is possessed by obligate anaerobes representative of seven genera; for this reason it has been proposed that these dissimilatory sulphate-reducing bacteria be treated taxonomically as a single physiological–ecological group. They include *Desulfovibrio, Desulfotomaculum, Desulfobacter, Desulfobulbus, Desulfococcus, Desulfonema* and *Desulfosarcina* (Postgate, 1984); of these, the first four genera are chemoheterotrophs while the last three are chemoautotrophs capable of growth on formate plus sulphate without the presence of organic substrates, and *Desulfonema* and *Desulfosarcina* additionally are able to grow on hydrogen and carbon dioxide plus sulphate. These organisms are of considerable ecological and economic importance. They are generally found in muddy environments where their conversion of sulphate to hydrogen sulphide is not only responsible for the deposition of metal sulphide ores but is also malodorous and causes damage to metals, particularly corrosion of buried gas and water pipes.

Mechanism of sulphate reduction. The reduction of sulphate to sulphide occurs via a sequence of three reactions catalysed respectively by *ATP sulphurylase* (1), *adenosine phosphosulphate* (*APS*) *reductase* (2), and a *bisulphite reductase* (3), enzymes that are common to all organisms capable of dissimilatory sulphate reduction but which are not unique to sulphate reducing bacteria. *Inorganic pyrophosphatase* (4) hydrolyses the pyrophosphate produced in reaction (1) and thereby ensures that this reversible reaction proceeds from left to right.

$$ATP + SO_4^{2-} \rightleftharpoons APS + PP_i \tag{1}$$
$$APS + 2e^- \rightleftharpoons AMP + SO_3^{2-} \tag{2}$$
$$HSO_3^- + 6H^+ + 6e^- \rightarrow HS^- + 3H_2O \tag{3}$$
$$PP_i + H_2O \rightarrow 2P_i \tag{4}$$

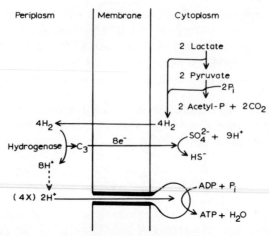

Figure 8.3. The hydrogen cycling mechanism proposed for electron transfer phosphorylation in *Desulfovibrio* growing on lactate plus sulphate. Abbreviations: c_3, cytochrome c_3; acetyl-P, acetyl phosphate.

Thus the reduction of sulphate effectively expends two ATP equivalents and to account for the net production of ATP during growth anaerobic electron-transfer phosphorylation has been invoked.

ATP sulphurylase primes sulphate to APS with the expenditure of one ATP, thereby raising the extremely low redox potential of the SO_4^{2-}/SO_3^{2-} couple ($E_m = -480$ mV at pH 7) by about 420 mV to that of the $APS/(AMP + SO_3^{2-})$ couple and thus enabling sulphate, in the form of adenosine phosphosulphate, to act as an oxidant. APS reductase is a non-haem iron-sulphur flavoprotein ($M \simeq 220k$) which contains 1 molecule of FAD and 12 atoms each of iron and acid-labile sulphur.

Three types of bisulphite reductase (which appear to be randomly distributed in the seven genera) have been recognized, namely desulfoviridin, pigment P_{582} and desulforubidin. Each is a haemoporphyrin protein with sirohaem or siroporphyrin (P_{582}) as prosthetic group. The enzymes are $\alpha_2\beta_2$ tetramers with $M \simeq 200k$ and possess four 4Fe–4S clusters and two haem groups per molecule. They display an optimum pH of about 6, at which value SO_3^{2-} is protonated (HSO_3^-). The overall six-electron transfer reaction shown in the equation above most probably occurs *in vivo* although three successive two-electron transfers, yielding trithionate ($S_3O_6^{2-}$) and thiosulphate ($S_2O_3^{2-}$) as intermediates, can occur in *in-vitro* assays; moreover, *Desulfovibrio* contains a very active thiosulphate reductase.

Carbon metabolism and energetics. Reducing equivalents for sulphate reduction and carbon for growth of the chemoheterotrophic organisms are secured from lactate or pyruvate, or from hydrogen as reductant with acetate as carbon source. The overall reaction with lactate in *Desulfovibrio* is

$$2\,\text{lactate} + SO_4^{2-} \rightarrow 2\,\text{acetate} + 2CO_2 + S^{2-}$$

The presence of lactate dehydrogenase, pyruvate:ferredoxin oxidoreductase, phosphotransacetylase and acetate kinase permits the conversion of lactate or pyruvate to acetate and CO_2 (and also H_2 although it is not usually a major product) with the conservation of 1 ATP by substrate-level phosphorylation. As these substrates yield $4e^-$ and $2e^-$ respectively, while the reduction of sulphate to sulphide in *Desulfovibrio* requires $8e^-$ and 2ATP equivalents, it will be apparent that substrate-level phosphorylation alone cannot supply the net energy yield needed for growth and must therefore be supplemented by electron-transfer phosphorylation.

However, a major difference exists with bacteria of the genus *Desulfotomaculum* which do not display significant inorganic pyrophosphatase activity but possess acetate-pyrophosphate kinase, catalysing the reaction

$$\text{acetate} + PP_i \rightarrow \text{acetyl phosphate} + P_i$$

thereby permitting conservation of the high energy function of pyrophosphate and ATP synthesis via acetate kinase

$$\text{acetyl phosphate} + ADP \rightarrow \text{acetate} + ATP$$

Consequently *Desulfotomaculum* conserve 2ATP per lactate oxidized and thus during growth on lactate plus sulphate have the potential for producing one net ATP per sulphate ion reduced.

$$2\,\text{lactate} + SO_4^{2-} + ADP + P_i \rightarrow 2\,\text{acetate} + 2CO_2 + S^{2-} + ATP$$

Unlike *Desulfovibrio*, therefore, electron-transfer phosphorylation is not mandatory for their growth with organic substrates and sulphate, and indeed does not occur, a conclusion supported by relative growth yield studies which show that yields of *Desulfotomaculum* are always lower than those of *Desulfovibrio*.

Acetate has been shown to be oxidized anaerobically to CO_2 in *Desulfovibrio postgatei* by a conventional tricarboxylic acid cycle, with the reductive synthesis of pyruvate from acetyl-SCoA and CO_2 serving as an anaplerotic reaction (p. 45).

Gaseous hydrogen is involved in the carbon metabolism of *Desulfovibrio* via the action of pyruvate:ferredoxin oxidoreductase, formate dismutase and hydrogenase. Reducing power from these sources and from lactate dehydrogenase is linked to the APS and bisulphite reductases by a low redox potential electron-transfer chain which comprises ferredoxin, flavodoxin, menaquinone, cytochrome b and, in *Desulfovibrio* but not in *Desulfotomaculum*, cytochrome c_3. Flavodoxins are flavoproteins ($M \simeq$ 16k) with FAD as prosthetic group; they can substitute for ferredoxin which performs a dual function in these organisms by accepting reducing equivalents from pyruvate and subsequently in the electron-transfer chain donating them to the APS and bisulphite reductases. This double site of operation is possible because both of these electron carriers can function at two different redox potentials which, in the case of ferredoxin, depends upon its state of aggregation, e.g. as a trimer (Fd I) it displays an E_m of $-440\,\text{mV}$ while in tetrameric form (Fd II) its E_m is $-140\,\text{mV}$; flavodoxin can fluctuate between its quinone and semiquinone forms ($E_m = -150\,\text{mV}$) and semiquinone and hydroquinone forms ($E_m = -440\,\text{mV}$). Cytochrome c_3 ($M = 13\text{k}$) was first discovered in *Desulfovibrio* and has the lowest known potential of any cytochrome ($E'_0 = -205$ to $-260\,\text{mV}$, pH 7). It has four haem c groups per molecule each exhibiting a different E_m value, which accounts for the range of E'_0 quoted, and is believed to accept electrons from hydrogenase and ferredoxin II. Also present in sulphate-reducing bacteria are cytochrome c_{553} ($E_m = +50\,\text{mV}$) and rubredoxin ($M = 6\text{k}$, $E_m = -60\,\text{mV}$, with a single iron atom and devoid of labile sulphur) but their function in electron transport remains speculative.

Energy transduction. Energy transduction presents different problems in *Desulfotomaculum* and *Desulfovibrio* for, as we have noted, the former genus effects substrate-level phosphorylation at the expense of inorganic pyrophosphate by virtue of its possession of acetate-pyrophosphate kinase, an enzyme absent from the genus *Desulfovibrio* which dissipates the high energy function of pyrophosphate by hydrolysis. In consequence growth of *Desulfotomaculum* can be supported by substrate-level phosphorylation alone whereas *Desulfovibrio* additionally requires electron-transfer phosphorylation, one proposal for which is the *hydrogen-cycling mechanism* (Peck, 1984).

First, however, in order to appreciate this proposed mechanism, the disposition of the redox carriers in sulphate respiration must be considered. Hydrogenase and cytochrome c_3 are located in the periplasm while ferredoxin, flavodoxin, APS-, bisulphite- and pyruvate:ferredoxin-reductases are cytoplasmic; only lactate dehydrogenase, menaquinone and cytochromes b and c are membrane-bound. The hydrogen cycling mechanism envisages that for each sulphate ion reduced to sulphide two molecules of lactate are oxidized in the cytoplasm, thus

$$2\,\text{lactate} \rightarrow 2\,\text{acetate} + 2CO_2 + 4H_2$$

Hydrogen, being a permeant molecule, diffuses across the cytoplasmic membrane to the periplasm where it is oxidized by hydrogenase $(4H_2 \rightleftharpoons 8H^+ + 8e^-)$ with cytochrome c_3 as cofactor. Only the eight electrons are transferred back across the membrane to the cytoplasm where they are used to reduce APS to sulphide with the consumption of eight protons. ATP required for the ATP sulphurylase reaction is, as previously noted, derived from acetyl phosphate. The net effect of this sequence is to generate a proton gradient across the membrane without the operation of a typical redox loop; eight protons are produced in the periplasm by the oxidation of $4H_2$ and eight protons are consumed in the cytoplasm by the reduction of sulphate to sulphide. ATP is synthesized via a proton-translocating ATPase or, alternatively, the proton gradient can be used to drive transport processes. Production of protons coupled to the oxidation of hydrogen has been reported with a $\rightarrow H/2e^-$ quotient of 2.0.

Support for the concept of hydrogen cycling comes from the observation that spheroplasts of *Desulfovibrio gigas*, which have lost all their periplasmic hydrogenase and cytochrome c_3, are unable to oxidize lactate with sulphate; the addition of purified hydrogenase plus cytochrome c_3 restores their lactate oxidizing ability. Implicit in the hydrogen cycling mechanism is that reducing equivalents derived from organic substrates in the cytoplasm be not used directly for sulphate reduction, a restriction possibly achieved by the use of different electron carriers, different specificities of carriers, compartmentation of the electron transfer systems or regulation of the dehydrogenase and reductases. If sulphate is absent, the hydrogen diffuses from the periplasm to the environment where methanogenic bacteria can utilize it for the reduction of CO_2 to methane.

CHAPTER NINE

ENERGY CONSERVATION IN
CHEMOLITHOTROPHIC BACTERIA

9.1 Characteristics of chemolithotrophy

The unique characteristics of the chemolithotrophic bacteria reside in their abilities (1) to secure all the energy they require for growth from the oxidation of inorganic compounds, and (2) with very few exceptions, to synthesize their cellular components solely from carbon dioxide, i.e. they are autotrophs. Consequently they are able to colonize entirely inorganic aqueous habitats which furnish their requisite reductant, a nitrogen source, carbon dioxide and essential mineral nutrients. Chemolithotrophs thus differ from photolithotrophs (Chapter 10), which use radiant energy and obtain reducing equivalents from the oxidation of inorganic sulphur compounds or hydrogen, in obtaining the ATP and NAD(P)H needed for biosynthesis solely from the oxidation of substances such as iron, sulphur, hydrogen, ammonia, nitrite and sulphur compounds (Figure 9.1). The terminal electron acceptor in most cases is molecular oxygen but a few species can grow anaerobically by replacing oxygen with nitrate. Some chemolithotrophs are able to assimilate organic substrates, possibly at the

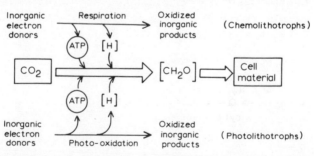

Figure 9.1. Fundamental reactions of autotrophy: comparison of chemolithotrophy and photolithotrophy. After Kelly (1978).

expense of chemolithotrophically-derived energy, and may in their natural environment display the phenomenon of *mixotrophy*, i.e. carbon is obtained from both CO_2 and organic compounds and energy from inorganic oxidations. The majority of their cell substance, however, is secured by the reductive assimilation of CO_2 via the Calvin cycle and the presence of the two key enzymes of this metabolic pathway, phosphoribulokinase and ribulose 1,5-bisphosphate carboxylase, distinguishes the chemolithotrophs from heterotrophic bacteria. The reactions of the Calvin cycle are described in a later section (9.3).

9.2 Energy transduction: reversed electron transfer

Energy transduction in chemolithotrophs occurs principally by oxidative phosphorylation involving an electron transfer chain of conventional type (Figure 9.2). A few *Thiobacillus* species, e.g. *T. thioparus, T. denitrificans* and *T. ferrooxidans*, additionally employ substrate-level phosphorylation during chemolithotrophic growth on inorganic sulphur compounds. Besides ATP, the reducing power needed for CO_2 assimilation is provided by NADH generated by the electron transfer chain. However, inspection of Figure 9.2 reveals an immediate problem, namely that, with the exception of molecular hydrogen, all the inorganic reductants channel electrons into the respiratory chain at potentials significantly higher than that of the $NAD^+/NADH$ couple, and they are therefore unable to reduce NAD^+ by forward electron transfer. Reduction of NAD^+ can only be achieved by energy-dependent *reversed electron transfer*, the precise energy requirement depending upon the point of electron entry to the chain. Again by reference to Figure 9.2, it will be apparent that a substrate which, for example, reduces cytochrome *c*, can by forward electron transfer to oxygen generate ATP at site 3, whereas reverse electron flow from cytochrome *c* to NAD^+ entails successive reversal of oxidative phosphorylation at sites 2 and 1, each involving the expenditure of energy. Thus, with the exception of

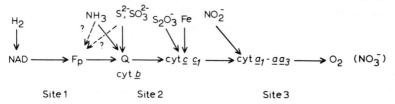

Figure 9.2.   Chemolithotrophic respiratory chain indicating location of electron entry from oxidation of inorganic substrates.

molecular hydrogen, a variable proportion of the energy derived from the oxidation of inorganic substrates is employed to drive NAD^+ reduction via reversed electron transfer. Specific examples will be considered later. Forward and reverse electron transfers are controlled by the adenylate energy charge and the $NADH/NAD^+$ ratio of the cell.

9.3 Carbon dioxide fixation: the Calvin cycle

Common to all autotrophs is the fixation of carbon dioxide and its reduction to the level of carbohydrate by a series of reactions which constitutes the reductive pentose phosphate pathway or Calvin cycle. The primary product of the fixation reaction is 3-phosphoglyceric acid which arises in a reaction between CO_2 and ribulose 1,5-bisphosphate catalysed by *ribulose bisphosphate carboxylase*.

$$
\begin{array}{ccc}
\begin{array}{l}
CH_2OPO_3^{2-} \\
| \\
C{=}O \\
| \\
H{-}C{-}OH \\
| \\
H{-}C{-}OH \\
| \\
CH_2OPO_3^{2-}
\end{array}
& + \quad CO_2 \quad \longrightarrow &
\begin{array}{l}
CH_2OPO_3^{2-} \\
| \\
HO{-}C{-}H \\
| \\
{}^-O{-}C{=}O \\
\overset{+}{} \\
{}^-O{-}C{=}O \\
| \\
H{-}C{-}OH \\
| \\
CH_2OPO_3^{2-}
\end{array} \\
\text{ribulose 1,5-bisphosphate} & & \text{3-phosphoglycerate}
\end{array}
$$

The substrate for fixation is formed by the phosphorylation of ribulose 5-phosphate, an intermediate of the pentose phosphate pathway (p. 31), by the action of *ribulose 5-phosphate kinase* (*phosphoribulokinase*), thus:

$$\text{ribulose 5-phosphate} + \text{ATP} \rightarrow \text{ribulose 1,5-bisphosphate} + \text{ADP}$$

The kinase and carboxylase are the key enzymes of autotrophy and are therefore characteristic of both chemolithotrophic and photoautotrophic bacteria. The remainder of the reactions which comprise the Calvin cycle (Figure 9.3) are catalysed by enzymes associated with the pentose phosphate and gluconeogenic pathways.

The conversion of 3-phosphoglycerate to triose and hexose sugars proceeds via its reduction to 3-phosphoglyceraldehyde which then enters the gluconeogenic pathway to yield fructose 6-phosphate (via triose phosphate isomerase, fructose bisphosphate aldolase and fructose 1,6-bisphosphatase). This fructose 6-phosphate then enters the pentose phos-

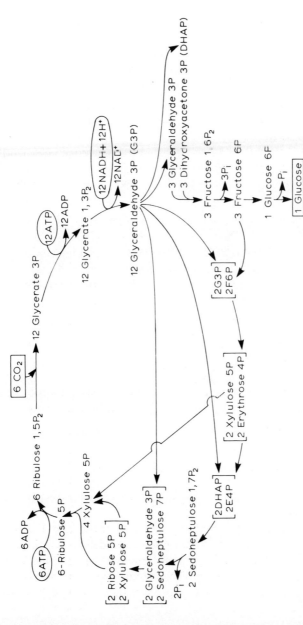

Figure 9.3. The reductive pentose phosphate or Calvin cycle for the conversion of CO_2 into glucose. For each turn of the cycle 1 mole of hexose is synthesized from $6CO_2$ at the expense of 18ATP and 12NADH, this energy being derived either from light (photoautotrophs) or from the oxidation of inorganic substrates (chemolithotrophs).

phate pathway in a reaction with 3-phosphoglyceraldehyde, catalysed by transketolase, giving xylulose 5-phosphate and erythrose 4-phosphate. Another molecule of 3-phosphoglyceraldehyde then reacts with the erythrose 4-phosphate, under the influence of aldolase, yielding sedoheptulose 1,7-bisphosphate. The phosphate in the 1-position of this sugar is removed by the action of sedoheptulose 1,7-bisphosphatase and the resulting sedoheptulose 7-phosphate reacts with a further molecule of 3-phosphoglyceraldehyde, again catalysed by transketolase, to produce ribose 5-phosphate and xylulose 5-phosphate. The pentose phosphates produced in this series of reactions are then isomerized (ribose 5-phosphate) or epimerized (xylulose 5-phosphate) to ribulose 5-phosphate, the substrate for the kinase, thereby completing the cycle. Energy consumption is high in terms of both ATP and reducing equivalents, as the following equation for the overall synthesis of one molecule of fructose 6-phosphate, by one turn of the cycle, reveals

$$6\,CO_2 + 6\text{ ribulose 1,5-bisphosphate} + 18\,ATP + 12\,NAD(P)H + 12\,H^+ + 2\,H_2O \longrightarrow$$
$$\text{fructose 6-phosphate} + 6\text{ ribulose 1,5-bisphosphate} + 18\,ADP + 17\,P_i + 12\,NAD(P)^+$$

Thus 18 ATP and 12 NAD(P)H are required for each molecule of hexose synthesized from $6\,CO_2$. This represents the major energy demand imposed upon chemolithotrophs and photoautotrophs and may account for at least 80% of their total biosynthetic energy requirement.

The energy-requiring steps are therefore (1) the phosphorylation by ATP of ribulose 5-phosphate to ribulose 1,5-bisphosphate and of 3-phosphoglycerate to 1,3-bisphosphoglycerate, and (2) the reduction of the last compound to 3-phosphoglyceraldehyde and P_i which accounts for the energy needed in the form of reducing equivalents. Of the sequence of reactions involved in the regeneration of ribulose 5-phosphate, only the enzyme sedoheptulose 1,7-bisphosphatase is not also encountered in the pentose phosphate oxidative pathway.

9.4 The Nitrobacter: oxidation of nitrite to nitrate

The Nitrobacter are obligately aerobic soil organisms which carry out the overall reaction

$$NO_2^- + \tfrac{1}{2}O_2 \rightarrow NO_3^- \qquad \Delta G^{0'} = -74.8\,kJ\,mol^{-1}$$

Isotopic experiments have shown that the oxygen atom incorporated in nitrate is derived from water and not from molecular oxygen, e.g.

$$NO_2^- + H_2^{18}O + \tfrac{1}{2}O_2 \rightarrow N^{18}O_3^- + H_2O,$$

thus it is reducing equivalents derived from this water that flow by the respiratory chain to oxygen. These organisms contain at least five cytochromes, namely cytochrome $c(E_m = +274\,\text{mV})$, cytochrome $a(E_m = +240\,\text{mV})$, cytochrome $a_3(E_m = +400\,\text{mV})$ and two cytochromes a_1 (with $E_m = +100\,\text{mV}$ and $+352\,\text{mV}$ respectively). There are also several iron–sulphur proteins with high E_m values and a molybdenum centre ($E_m = +340\,\text{mV}$). Invaginations of the cytoplasmic membrane, resembling mitochondrial cristae, occur in these organisms, and vesicles have been widely used to investigate the efficiency of energy coupling in nitrite oxidation. Experiments with reagents which specifically dissipate the ΔpH and $\Delta\psi$ components of the protonmotive force of vesicles permitted the conclusion that the rate of nitrite oxidation is directly related to $\Delta\psi$ and not inversely related to ΔpH. Thus NH_4^+ and certain amines which collapse ΔpH by electroneutral mechanisms (and thereby enhance the $\Delta\psi$ component of Δp), stimulated nitrite oxidation and the reduction of cytochrome c but did not affect NADH oxidation, whereas valinomycin plus K^+, reagents which collapse $\Delta\psi$ and enhance ΔpH, decreased the rates of nitrite oxidation and reduction of cytochrome c, leaving NADH oxidation unaffected. However, although NADH oxidation was not influenced by reagents which specifically collapse either ΔpH or $\Delta\psi$, it was subject to respiratory control, and could be released by $ADP + P_i$ or by carbonyl cyanide-p-trifluoromethoxyphenylhydrazone (FCCP).

The foregoing evidence supports the belief that while the overall oxidation of nitrite generates a Δp to drive ATP synthesis, the nitrite–cytochrome c reductase step is energy-dependent. Cobley (1976) proposed a chemiosmotic model (Figure 9.4) to account for these observations. Essentially oxidation of nitrite by cytochrome a_1/Mo is accompanied by the energetically 'uphill' vectorial transfer of a hydride ion to cytochrome c, with extrusion of a proton to the exterior. Two e^- are transferred back to cytochrome oxidase aa_3 and react with molecular oxygen completing the loop. The $\Delta\psi$ component of the Δp generated by the operation of the loop drives the initial H^- transfer. The scheme is supported by the fact that KBH_4 mimics nitrite reduction of cytochrome, and by the membrane orientation of the respiratory components and their redox potentials. The scheme predicts a $\rightarrow H^+/O$ quotient of unity, but experimental verification is difficult because nitrite oxidation is inhibited under conditions which maximize the contribution of ΔpH to Δp.

Inside-out vesicles which displayed very high efficiencies of ATP synthesis have been prepared, e.g. ATP/O quotients of up to 3 for NADH oxidation and ATP/NO_2^- quotients of 0.5 to 0.9 for NO_2^- oxidation, the last

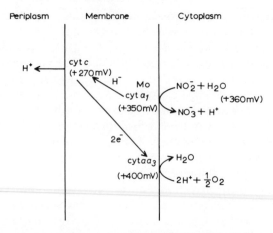

Figure 9.4. Chemiosmotic scheme to explain H^+ translocation in *Nitrobacter winogradskyi* oxidizing nitrite with oxygen (adapted from Cobley, 1976). The reduction of cytochrome *c* by H^- is driven by $\Delta\psi$ derived from the entire loop. An associated proton-translocating ATP synthetase would permit ATP synthesis with the H^+/ATP stoichiometry of 2, i.e. ATP/NO_2^- stochiometry of 0.5. The redox potentials shown are E_m values.

value (0.9), however, being too high to be accommodated by the value of 0.5 predicted from the $\rightarrow H^+/O$ quotient of unity (Figure 9.4). The problem is unresolved and the present tendency has been to accept an ATP/NO_2^- quotient of unity.

Energy expenditure for reversed electron transfer from nitrite to NAD^+ involves nitrite–cytochrome *c* reductase and sites 1 and 2 and is therefore expensive in terms of ATP. A theoretical minimum of 3 ATP is necessary but indications are that as many as 5 ATP (equivalent to 5 NO_2^- oxidized) are needed for the reduction of one molecule of NAD^+. When the reducing equivalents required for CO_2 assimilation are also taken into account, an overall energy requirement for the reduction of CO_2 can be deduced. Thus, to assimilate CO_2 via the Calvin cycle (p. 113), 3 ATP and 2 NADH are required per molecule. Accepting that 5 ATP are required per NAD^+ reduced, then a total of 13 ATP will be utilized, equivalent to 13 NO_2^- oxidized. Additionally, however, reducing equivalents are needed for the reduction of the 2 NAD^+, i.e. two further NO_2^- must be oxidized to generate 2 NADH, making a total of 15 NO_3^- oxidized per CO_2 reduced to $[CH_2O]$. These heavy demands on energy generation are reflected by the very low growth yields of *Nitrobacter*.

9.5 *Nitrosomonas* species: the oxidation of ammonia to nitrite

The oxidation of ammonia to nitrite by soil bacteria of the genus *Nitrosomonas* is represented by the equation

$$NH_3 + 1\tfrac{1}{2}O_2 \rightarrow NO_2^- + H^+ + H_2O \quad \Delta G^{0\prime} = -272\,kJ\,mol^{-1}$$

and hydroxylamine (NH_2OH) and nitroxyl (NOH) are believed to be intermediates. The redox potential of the NH_3/NH_2OH couple is so electropositive ($E_0' = +899\,mV$) that none of the components of the respiratory chain can be directly reduced by ammonia and, moreover, the reaction has an unfavourable equilibrium constant.

$$NH_3 + \tfrac{1}{2}O_2 \rightarrow NH_2OH \quad \Delta G^{0\prime} = +16\,kJ\,mol^{-1}$$

It has been suggested that the reaction might be achieved by an oxygenation involving the enzyme ammonia hydroxylase and a cytoplasmic *b*-type cytochrome designated P460

$$NH_3 + O_2 + XH_2 \rightarrow NH_2OH + H_2O + X$$

where XH_2 is possibly equivalent to $2(P460-Fe^{2+} + H^+)$. The next step is the oxidation of hydroxylamine to the highly unstable nitroxyl, effected by hydroxylamine-cytochrome *c* reductase and the cytochrome system, followed by oxidation of nitroxyl to nitrite via a second cytochrome P460-dependent oxygenation reaction

$$NH_2OH + 2\,cyt\,c\text{-}Fe^{3+} \rightarrow [NOH] + 2\,cyt\,c\text{-}Fe^{2+} + 2H^+$$
$$[NOH] + 2\,P460\text{-}Fe^{3+} + H_2O \rightarrow NO_2^- + H^+ + 2\,P460\text{-}Fe^{2+} + 2H^+$$

As the cytochrome P460-mediated reactions occur in the cytoplasm, proton translocation is possibly achieved via a redox loop comprising hydroxylamine cytochrome *c* reductase and cytochrome oxidase, but details of the membrane organization are not available.

Experiments suggest there is only one site involved in proton translocation with either ammonia or hydroxylamine as the electron donor, and ATP synthesis occurs only at site 3, i.e. $ATP/NH_3 = 1$. NAD^+ reduction is achieved by reversed electron flow from hydroxylamine or cytochrome $c\text{-}Fe^{2+}$ with concomitant energy expenditure at sites 1 and 2. Consequently molar growth yields for *Nitrosomonas* species are very low.

9.6 *Thiobacillus* species: the sulphur-oxidizing bacteria

The pathways of metabolism of sulphur compounds in the Thiobacilli still hold some uncertainties despite considerable research. These organisms

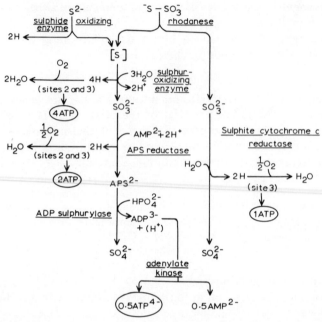

Figure 9.5. Scheme for oxidation of sulphide and thiosulphate by Thiobacilli. After Kelly (1978) and Jones (1982).

cleave thiosulphate to sulphite and sulphane–sulphur [S] by the action of the enzyme rhodanese prior to oxidation. The [S] is oxidized to sulphite by an iron–sulphur flavoprotein and electrons traverse the respiratory chain to oxygen via coupling sites 2 and 3 (Figure 9.5). The subsequent oxidation of sulphite to sulphate occurs by two pathways, one involving direct oxidation catalysed by a sulphite-cytochrome c reductase with electron transfer to oxygen via cytochromes o and site 3, and the other occurring via the adenosine 5′-phosphosulphate (APS, Figure 9.6) route with substrate-level phosphorylation (APS is a high energy compound with $\Delta G^{\circ\prime}$ of hydrolysis of $-88\,\mathrm{kJ\,mol^{-1}}$). This latter pathway entails the iron–sulphur

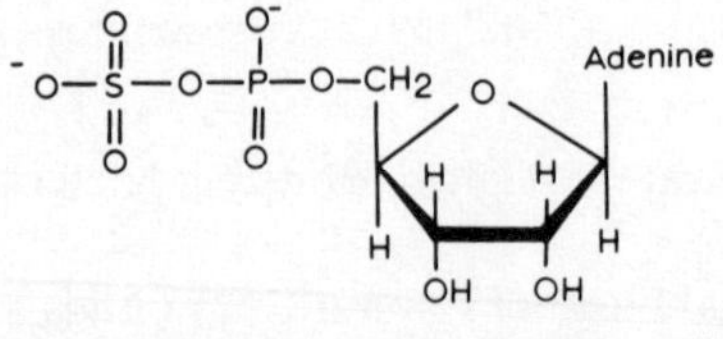

Figure 9.6. Adenosine 5′-phosphosulphate (APS).

flavoprotein APS reductase, ADP sulphurylase and adenylate kinase in the following sequence of reactions:

$$2SO_3^{2-} + 2AMP \rightarrow 2APS + 4e^-$$
$$2APS + 2P_i \rightarrow 2ADP + 2SO_4^{2-}$$
$$2ADP \rightleftharpoons ATP + AMP$$

and thus 0.5 ATP is generated via substrate-level phosphorylation per sulphite ion oxidized. The $4e^-$ enter the respiratory chain at the level of flavoprotein, and their transfer to oxygen therefore encompasses sites 2 and 3 with 4ATP generated or $2ATP/SO_3^{2-}$. Consequently, depending upon the proportion of sulphite entering the two pathways, the aerobic oxidation of thiosulphate can generate between 6 and 9 ATP per mol. Some Thiobacilli are able to circumvent the ADP sulphurylase and adenylate kinase steps by an ATP sulphurylase which reacts APS directly with inorganic pyrophosphate,

$$APS^{2-} + PP_i \rightarrow ATP + SO_4^{2-}$$

thus doubling the ATP yield derived from substrate-level phosphorylation.

All the Thiobacilli employ energy-dependent reversed electron transfer from thiosulphate or sulphite to reduce the NAD^+ needed for biosynthetic purposes and their molar growth yields are therefore low, as would be expected. They are usually able to grow anaerobically by replacing oxygen with nitrate as the terminal electron acceptor.

9.7 *Thiobacillus ferro-oxidans*: the oxidation of ferrous iron

Thiobacillus ferro-oxidans is an acidophile which can grow at pH values of about 2 and uses the oxidation of Fe^{2+} to Fe^{3+} by oxygen as the sole source of energy for growth (Ingledew, 1982; Cox and Brand, 1984). In this organism the pH differential of the cytoplasm and external medium provides the energy for ATP synthesis by driving a proton-translocating ATP synthetase, a transmembrane ΔpH of 4.5 units being equivalent to about 270 mV. Accumulation of these translocated protons in the cyto-plasm is prevented by the operation of the respiratory chain which possesses several c-type cytochromes, cytochrome oxidase a_1 (containing two a haems with $E_m + 420$ mV and $+ 500$ mV respectively), and rusticyanin, a copper-containing protein with $E_m = + 680$ mV. The rusticyanin and one of the cytochromes c are in the periplasm while another cytochrome c is on the periplasmic surface of the membrane and the O_2-binding site of cytochrome oxidase is on the cytoplasmic side (Figure 9.7). This vectorial

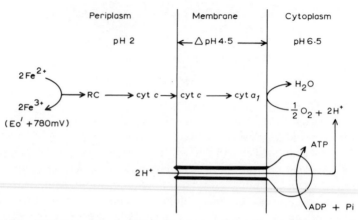

Figure 9.7. Oxidation of iron by *Thiobacillus ferro-oxidans* with concomitant cytoplasmic uptake of protons. RC, rusticyanin. After Ingledew *et al.* (1977).

arrangement allows electron transfer from Fe^{2+} to oxygen with the concomitant removal of protons from the cytoplasm to form water.

The redox components needed for reversed electron transfer from Fe^{2+} to NAD^+ remain to be established but presumably three coupling sites must be traversed, indicating that at least 6 Fe^{2+} must be oxidized per NADH formed. Very low growth yields (e.g. 1.33 g per mole Fe^{2+} oxidized) are therefore manifest by *T. ferro-oxidans*.

9.8 The autotrophic methanogens: reduction of carbon dioxide to methane

Some micro-organisms are able to utilize and produce C_1 compounds more reduced than CO_2, of which the principal examples are methane and methanol. The carbon assimilation route in these organisms is quite different from that observed in aerobic methylotrophs (p. 123). The Calvin cycle is not involved and cell constituents are produced from a *de novo* synthesis of acetyl-SCoA from CO_2, a route also operating in those anaerobes that ferment glucose to three molecules of acetate (homoacetogenic bacteria).

Methanogens are obligate anaerobes that carry out the reduction of carbon dioxide to methane using molecular hydrogen as the reductant.

$$CO_2 + 4H_2 \rightarrow CH_4 + 2H_2O \quad \Delta G^{o'} = -135.6 \, kJ \, mol^{-1}$$

These organisms are found in lake sediments and the animal rumen where they associate with cellulolytic hydrogen-producing bacteria. They are now classified as Archaebacteria (together with the extreme halophiles and thermoacidophiles) on account of their distinctive composition, which includes, for example, the absence of peptidoglycan from their cell walls and the unique presence of coenzyme M (2-mercaptoethanesulphonic acid) and coenzyme F_{420}, a modified flavin (FMN) derivative.

The anaerobic electron transfer system which operates in methane formation seems to be unique to methanogenic bacteria. It is sensitive to oxygen and its components are believed to be membrane-bound although much remains to be discovered about this system. The initial reaction involves a membrane-associated, nickel-containing hydrogenase which effects the reaction

$$H_2 \rightleftharpoons 2H^+ + 2e^- \qquad E'_0 = -414\,mV$$

The sequence of operation of other components of the redox chain is not defined: they include coenzyme M ($E'_0 = -193\,mV$), coenzyme F_{420} ($E'_0 = -373\,mV$), methanopterin ($E'_0 = -450\,mV$) and other factors designated F_{342} and F_{430}. It is possible that the hydrogenase, methanopterin and F_{420} function in respiration from H_2 to $NADP^+$.

The reduction of CO_2 does not involve methanol, formaldehyde or formate as free intermediates (although these compounds can serve as precursors of methane) and the methyl derivative of coenzyme M ($CH_3.CoM$) has been identified as the substrate for *methyl-coenzyme M reductase* ($M = 300\,k$) which catalyses the final step in methane production

$$CH_3.CoM + H_2 \xrightarrow{ATP,Mg^{2+}} CH_4 + HS.CoM$$

However, the reaction is also dependent upon hydrogenase and a cofactor. The ATP requirement is catalytic and 1 mol of ATP stimulates the formation of 5 to 15 mol of methane, possibly by phosphorylating or adenylating one of the enzymes of the methanogenic system.

Recent work has demonstrated that when *Methanobacterium thermoautotrophicum* grows on H_2/CO_2, the methane hydrogen atoms are derived exclusively from water and not from the molecular hydrogen, which thus serves solely as a source of electrons. Further, the initial stages of the reaction appear to involve reverse electron transfer while the terminal steps of methanogenesis are linked to forward electron flow, which provides the energy needed to drive the former. Studies with membrane vesicles have shown ATP synthesis to be driven by hydrogen oxidation (which generates a Δp with inside negative) or by a K^+ gradient, and to be inhibited by uncoupling reagents and by inhibitors of ATP synthetase. There is present in such vesicles an adenine nucleotide translocase, previously unknown in prokaryotes and which transports ATP and ADP into mitochondria; ATP synthesis is sensitive to inhibitors of this translocase. This unusual finding supports other evidence that the methanogenic bacteria probably contain discrete mini-organelles resembling mitochondria which are involved in energy transduction and methane formation.

The energy requirement for the synthesis of hexose from 6 CO_2 is 18 ATP and 12 NAD(P)H; the ATP must be derived from the reduction of CO_2 to methane while reducing equivalents can be secured from hydrogen via hydrogenase.

9.9 Hydrogen-oxidizing bacteria

Hydrogen-oxidizing organisms are all facultative chemolithotrophs and are usually classified with their chemoheterotrophic relatives. Examples include *Alcaligenes eutrophus*, *Pseudomonas saccharophila*, *Nocardia autotrophica* and *Paracoccus denitrificans*, the facultative phototroph *Rhodopseudomonas capsulata* and the chemoheterotrophic nitrogen-fixer *Azotobacter vinelandii*. These organisms are characterized by possession of the enzyme hydrogenase, of which both NAD$^+$-dependent and NAD$^+$-independent forms exist, the former being a cytoplasmic enzyme and the latter either periplasmic or membrane-bound. The NAD$^+$-enzyme is an iron–sulphur–FMN protein of $M \simeq 200$ k and the NAD$^+$-independent catalyst is a smaller iron–sulphur protein of $M \simeq 90$ k.

Experimental evidence points to the membrane-associated and periplasmic hydrogenase reactions being coupled to ATP synthesis at site 1 by virtue of their orientation in the membrane, which permits the uptake of 2H$^+$ from the cytoplasm and the extrusion of 2H$^+$ into the periplasm. Either the hydrogenase itself has a transmembrane location and usually binds hydrogen at the periplasmic surface, or a periplasmic enzyme is linked to a transmembrane electron transfer system, to effect the necessary protonmotive limb. The implication seems to be that reduction of NAD$^+$ for biosynthesis in these organisms is an energy-independent process, for it can occur either by a direct NAD$^+$-dependent hydrogenase or by reversed electron transfer at site 1 via NADH dehydrogenase, driven by the Δp generated by the membrane-bound NAD$^+$-independent hydrogenase.

9.10 Methylotrophic bacteria and yeasts

Methylotrophs are organisms that obtain their carbon and energy for growth by metabolizing reduced C_1 compounds as their sole source of carbon. In contrast to the obligately anaerobic methanogens previously discussed, with the exception of certain photosynthetic bacteria, they are aerobes. There are three principal pathways by which carbon is issimilated in these bacteria, and a fourth, different route exists in methylotrophic yeasts. Energy and reducing power for these assimilatory reactions are furnished by the dissimilatory pathways of oxidation of the growth substrates.

9.10.1 Assimilatory pathways

In bacteria these pathways include carbon assimilation via the Calvin cycle, the hexulose phosphate cycle or the serine pathway, while in yeasts the dihydroxyacetone pathway operates. The Calvin cycle has already been discussed (p. 112); the other pathways are found in heterotrophic methylotrophs which assimilate much of their growth substrate at the oxidation level of formaldehyde and fix CO_2 by reactions other than that catalysed by ribulose bisphosphate carboxylase. However, these pathways are also found in obligate methylotrophs.

The hexulose phosphate pathway. Formaldehyde is the reactant in a series of reactions somewhat analogous to those of the Calvin cycle although not, of course, requiring the reductive step. Thus formaldehyde reacts with ribulose 5-phosphate under the influence of *hexulose phosphate synthase* to form D-erythro-L-glycero-3-hexulose 6-phosphate, which is then isomerized to fructose 6-phosphate by *hexulose phosphate isomerase*:

$$
\begin{array}{ccc}
 & CH_2OH & CH_2OH \\
CH_2OH & HO-C-O & C=O \\
C=O & C=O & HO-C-H \\
H-C-OH+HCHO \rightarrow & H-C-OH & \rightleftharpoons \quad H-C-OH \\
H-C-OH & H-C-OH & H-C-OH \\
CH_2OPO_3{}^{2-} & CH_2OPO_3{}^{2-} & CH_2OPO_3{}^{2-} \\
\text{ribulose 5-phosphate} & \text{3-hexulose 6-phosphate} & \text{fructose 6-phosphate}
\end{array}
$$

The subsequent metabolism of fructose 6-phosphate occurs in two phases, namely *cleavage* of the hexose followed by *rearrangement* of the resulting triose phosphate with hexose phosphates to regenerate ribulose 5-phosphate.

The cleavage metabolism of fructose 6-phosphate occurs via two alternative routes, one involving the reactions of glycolysis to yield dihydroxyacetone phosphate and glyceraldehyde 3-phosphate, and the other conversion via glucose 6-phosphate, 6-phosphogluconate and the enzymes of the Entner–Doudoroff pathway to give pyruvate and glyceraldehyde 3-phosphate. Then follows a series of carbon rearrangements of triose phosphate and hexose phosphates to regenerate the ribulose 5-phosphate acceptor, in accordance with the overall equation

$$2\,\text{fructose 6-phosphate} + \text{glyceraldehyde 3 phosphate} \rightleftharpoons 3\,\text{ribulose 5-phosphate.}$$

Again, there are two possible routes by which this may be effected, one specifically requiring the operation of phosphofructokinase, fructose bisphosphate aldolase and sedoheptulose bisphosphatase, and the other needing transaldolase, while the enzymes transketolase, ribose 5-phosphate isomerase and ribulose 5-phosphate 3-epimerase are common to both pathways. It should be noted that the pathway involving phosphofructokinase entails the expenditure of 1 ATP to effect the $2C_6 + C_3 \rightarrow 3C_5$ conversion whereas the transaldolase route does not.

The formation of 1 molecule of hexose via the hexulose phosphate cycle may thus be represented by the overall equation

$$6\,\text{ribulose 5-phosphate} + 6\text{HCHO} \rightarrow 6\,\text{ribulose 5-phosphate} + C_6H_{12}O_6$$

The serine pathway. Unlike the hexulose phosphate route, the serine pathway does not effect assimilation of carbon exclusively as formaldehyde, about 30 to 50% being derived from CO_2 fixed by the phosphoenolpyruvate carboxylase reaction which forms part of this fairly complex metabolic sequence. The overall stoichiometry may be represented as

$$2\text{HCHO} + CO_2 + 3\text{ATP} + 2\text{NAD(P)H} + 2\text{H}^+ \rightarrow$$
$$3\text{-phosphoglycerate} + 2\text{NAD(P)}^+ + 2\text{H} + 3\text{ADP} + 2\,P_i + H_2O$$

where the 2H on the right-hand side derive from the succinate deydrogenase reaction. For full details of the metabolic sequences involved the reader is referred to the monographs by Anthony (1982) and Large (1983). Here it must suffice to point out that the pathway is so named because serine is the initial product detected when ^{14}C-labelled substrates such as methanol, formate or methylamine are metabolized. Three phases of metabolism are recognized in the pathway representing respectively (a) conversion of formaldehyde and CO_2 into acetyl-SCoA; (b) conversion of acetyl-SCoA into glycine via glyoxylate; and (c) net synthesis of 3-phosphoglycerate which serves as a C_3 precursor for cell biomass. Organisms using the serine pathway include *Pseudomonas* AM1, *Pseudomonas extorquens*, *Methanomonas methano-oxidans* and *Hyphomicrobium*.

The dihydroxyacetone pathway. Methylotrophic yeasts (species of *Candida*, *Pichia*, *Hansenula* and *Torulopsis*) assimilate carbon from methanol after its oxidation to formaldehyde. Isotopic incorporation patterns reveal sugar phosphates as initial products but the pathway differs from the hexulose phosphate route in reacting formaldehyde with xylulose 5-phosphate to yield dihydroxyacetone and glyceraldehyde 3-phosphate, under the influence of a novel type of transketolase enzyme called *dihydroxyacetone synthase*, thus

HCHO + xylulose 5-phosphate → dihydroxyacetone + glyceraldehyde 3-phosphate.

The dihydroxyacetone formed is then phosphorylated by a novel *triokinase* and the two triose phosphates are converted to fructose 1, 6-bisphosphate by fructose bisphosphate aldolase, and fructose 6-phosphate is derived by the action of fructose bisphosphatase, giving the following sequence:

dihydroxyacetone + ATP → dihydroxyacetone phosphate + ADP
dihydroxyacetone phosphate + glyceraldehyde 3-phosphate → fructose 1, 6-bisphosphate
fructose 1, 6-bisphosphate → fructose 6-phosphate + P_i

The overall stoichiometry per molecule of hexose synthesized is therefore

6HCHO + 6ATP → fructose 6-phosphate + 6ADP + $5P_i$

9.10.2 Dissimilatory pathways

The assimilation of reduced C_1 compounds makes energy demands on the methylotrophic cell which must be met by the generation of ATP in the dissimilatory reactions that effect the oxidation of these same compounds to CO_2. These oxidative reactions present some interesting features because many of the dehydrogenases employed are not $NAD(P)^+$-linked and their prosthetic groups and electron acceptors have not yet been fully elucidated. Experiments with redox dyes suggest that the natural electron acceptors will display more positive electrode potentials than the nicotinamide nucleotides, with the consequent implication that their reoxidation via the electron transfer chain will yield less ATP than does reoxidation of NAD(P)H. Further, certain of the dissimilatory processes employ monooxygenases (mixed function oxygenases or hydroxylases) consuming reducing power in reactions of the type

$$CH_4 + O_2 + NADPH + H^+ \rightarrow CH_3OH + NADP^+ + H_2O$$

where one atom of molecular oxygen is incorporated in the product and the other in the water molecule formed. Such reactions effectively decrease the availability of reducing power for oxidative phosphorylation and thus biosynthesis and growth require that the net production of reducing power exceeds that consumed in the mono-oxygenase reactions.

Methane is oxidized by the sequence of reactions

$$CH_4 \overset{(1)}{\rightarrow} CH_3OH \overset{(2)}{\rightarrow} HCHO \underset{(4)}{\overset{(3)}{\rightarrow}} HCOOH \overset{(5)}{\rightarrow} CO_2$$

in which reaction (1) is catalysed by a mono-oxygenase and reactions (2) and (3) by dehydrogenases not linked to $NAD(P)^+$. The bacterial methanol dehydrogenase, which effects reaction (2), contains a novel quinone prosthetic group called *methoxatin* or *pyrroloquinoline quinone* (PQQ), and is referred to as a quinoprotein. PQQ has also been identified in certain other bacterial dehydrogenases, e.g. glucose dehydrogenase (p. 36). Some four different types of methanol dehydrogenase have been discovered differing only slightly in specificity and molecular weight; they generally display activity with alkan-1-ols of chain lengths up to C_{11}, and with formaldehyde.

PQQ functions as a hydrogen carrier by virtue of its dione structure, undergoing reduction to the quinol form ($PQQ/PQQH_2$), and has an E_0' (pH 7) of about $+120\,mV$; reoxidation by the electron transfer chain occurs at the level of cytochrome c, emphasizing the point already made in relation to ATP generation by this sequence.

The oxidation of formaldehyde via reaction (3) is catalysed by the PQQ–methanol dehydrogenase. Oxidation via reaction (4) requires an NAD^+-dependent *formaldehyde dehydrogenase*, which also needs glutathione because the true substrate is S-hydroxymethylglutathione, and a second enzyme, *S-formylglutathione hydrolase*, to convert the product (S-formyl-glutathione) into formate. Some organisms possess only a third type of formaldehyde dehydrogenase, a haemoprotein enzyme.

Because negligible activities of formaldehyde dehydrogenase and formate dehydrogenase are detectable in many but not all micro-organisms employing the assimilatory hexulose phosphate cycle for growth on methanol, it has been proposed that this pathway in conjunction with 6-phosphogluconate dehydrogenase could function to oxidize formaldehyde in accordance with the equation

$$HCHO + 2NAD(P)^+ + H_2O \rightarrow CO_2 + 2NAD(P)H + 2H^+$$

thereby constituting a dissimilatory hexulose phosphate pathway. Such a pathway of formaldehyde oxidation obviously does not involve formate as an intermediate.

The oxidation of formate to CO_2 in most methylotrophs is effected by an NAD^+-dependent formate dehydrogenase but *Pseudomonas oxalaticus* also contains a membrane-bound enzyme which is independent of NAD^+.

The dissimilatory pathway for methanol in yeasts differs from that in bacteria by using an oxidase instead of a dehydrogenase. *Alcohol oxidase* is a non-covalently-bound FAD enzyme possessing eight identical subunits (each of $M \simeq 80$ k) with one FAD molecule associated with each. It possesses relatively low specificity for alkanols and alkenols but displays its highest affinity for methanol (K_m (app.) $= 0.2$ to 2 mM). It has a low affinity for oxygen (K_m (app.) $= 0.25$ to 0.4 mM) which effectively means the enzyme is always operating under non-saturating conditions because the usual concentration of oxygen in air-saturated buffer is only about 0.2 mM. The yeast cell offsets this low affinity for oxygen by synthesizing large amounts of the enzyme, often accounting for 10 to 20% of the soluble protein in extracts of methanol-grown yeasts. The reaction catalysed by alcohol oxidase is

$$CH_3OH + O_2 \rightarrow HCHO + H_2O_2$$

and the toxic hydrogen peroxide produced is decomposed to water and oxygen by catalase. Both the alcohol oxidase and catalase enzymes are located in organelles termed *peroxisomes* or *microbodies*, which are typical of most eukaryotic cells and also contain other oxidase enzymes that give rise to hydrogen peroxide (Veenhuis *et al.*, 1983). Yeasts grown on methanol contain many more, larger peroxisomes than do glucose-grown organisms and electron micrographs reveal that these organelles then occupy much of the cell volume.

Formaldehyde produced in the peroxisomes must pass into the cytosol for further oxidation. There the principal reaction is catalysed by an NAD^+–glutathione (GSH)-dependent formaldehyde dehydrogenase and, as already noted for the bacterial enzyme, the product is *S*-formylglutathione.

$$HCHO + GSH + NAD^+ \rightarrow HCO.SG + NADH + H^+$$

In some yeasts, e.g. *Candida boidinii*, a specific esterase hydrolyses the thiol ester to yield formate

$$HCO.SG + H_2O \rightarrow HCOOH + GSH$$

whereas in others, such as *Hansenula polymorpha*, HCO.SG is probably directly oxidized by an NAD^+-dependent formate dehydrogenase, the next enzyme in the sequence.

$$HCO.SG + NAD^+ + H_2O \rightarrow CO_2 + GSH + NADH + H^+$$

The K_m for formate is remarkably high (6 to 55 mM) whereas that for *S*-formylglutathione is much lower, which possibly accounts for the foregoing observation. However, some formaldehyde may also be oxidized by methanol oxidase (note that formaldehyde is extensively hydrated in solution).

One further point remains to be made with reference to dissimilatory pathways, namely that the tricarboxylic acid cycle does not fulfil any significant oxidative role during methylotrophic growth. Indeed, most organisms employing the hexulose phosphate assimilatory pathway appear to have an incomplete tricarboxylic acid cycle, lacking 2-oxoglutarate dehydrogenase, in which respect they therefore resemble facultative heterotrophs growing under anaerobic conditions (p. 44).

9.10.3 Energy transduction in methylotrophic bacteria

The dissimilatory oxidative reaction pathways previously discussed yield ATP to the organisms via the proton-translocating electron transfer chains. Considerable research has been carried out on the composition of the respiratory chains of a number of methylotrophs and while features common to those of other aerobic bacteria (Chapter 7) do exist, as might be expected, some variation is encountered as, for example, the PQQ-dependent methanol dehydrogenase which interacts with cytochrome *c*.

Generally, all methylotrophs grown on methane or methanol possess cytochromes *b* and *c* (the latter in very high concentrations) and usually cytochrome aa_3. Cytochromes *c*, of which there are several varieties, occupy a special role in methane and methanol metabolism. Measurements of E_0' (pH 7) for cytochromes *c* from different methylotrophs, e.g. *Pseudomonas* AM1, *Pseudomonas extorquens, Methylophilus methylotrophus*, give values of about 300 mV or higher, and the available evidence indicates that the PQQ-methanol dehydrogenase transfers its electrons directly to cytochrome *c*, thereby bypassing cytochrome *b*. Interestingly, with *Pseudomonas* AM1 it has been shown that cytochrome *c* is probably not normally involved in the oxidation of NADH or cytochrome b, except under conditions of carbon-limited growth on succinate.

Measurements of ATP synthesis with membrane vesicles have indicated that the oxidation of methanol to formaldehyde is coupled to the synthesis of ATP, the $P/2e^-$ quotient of about unity being lower than that observed with NADH and succinate; this suggests that cytochrome c is not involved in proton translocation in the cytochrome b/O_2 segment of the electron transfer chain. However, under carbon-limited conditions higher $\rightarrow H^+/O$ quotients were obtained corresponding to the operation of three proton-translocating segments, i.e. including cytochrome c.

A feature of considerable importance in relation to ATP generation via the proton translocating electron transfer chains of methylotrophs is the fact that they possess relatively few NAD^+-linked dehydrogenases, e.g. *Pseudomonas* AM1, which assimilates methanol via the serine pathway, has only formate dehydrogenase in this category. It is estimated that under these conditions less than 5% of electron transfer to oxygen is from NADH, some 50% from methanol dehydrogenase ($PQQH_2$), and the rest from formaldehyde dehydrogenase and flavoproteins. The implications for ATP synthesis are patent.

As we have seen, methanol oxidation to formaldehyde is associated with the formation of 1 ATP and the translocation of $2H^+$. The currently favoured arrangement of methanol dehydrogenase, cytochrome c and cytochrome aa_3 in the cytoplasmic membrane, to account for the experimental findings, envisages the PQQ-linked dehydrogenase and cytochrome c on the outer (periplasmic) side and cytochrome aa_3 on the inner side of the membrane. Thus in this scheme two protons are released from the $PQQH_2$ prosthetic group on the outside and do not actually traverse the membrane when the enzyme reacts and transfers electrons to cytochrome c. The pair of electrons then traverse the membrane from cytochrome c to aa_3 which catalyses reaction with oxygen in the cytosol to form water (Figure 9.8). The formaldehyde produced in the periplasm must subsequently enter the cell to participate in assimilatory reactions but the details remain obscure.

9.10.4 Energetics of C_1 assimilation sequences

We have noted that there are four possible variants of the hexulose phosphate assimilatory pathway involving permutations of the two cleavage and two rearrangement sequences. The energy balances of these variants differ and it is instructive to compare them one with the other and also

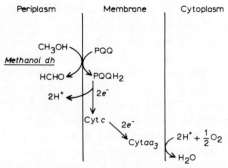

Figure 9.8. Proposed arrangement of methanol dehydrogenase (dh) and cytochromes *c* and *aa*₃ in the membrane of *Pseudomonas* AM1. After Anthony (1982).

with the balances for the Calvin cycle, serine pathway and dihydro-xyacetone route characteristic of yeasts. To simplify the comparison it is convenient to consider that triose phosphate and 3-phosphoglycerate are converted to pyruvate by glycolytic reactions involving the respective formation of (NADH + 2ATP) and ATP. The resulting energy balances are recorded in Table 9.1.

It will be apparent that two of the variants of the hexulose phosphate pathway for bacterial formaldehyde assimilation present no net ATP change and may be written as

$$3HCHO + NAD(P)^+ \rightarrow pyruvate + NAD(P)H + H^+$$

while that employing FBP aldolase and transaldolase achieves net synthesis of 1 ATP:

$$3HCHO + NADP^+ + ADP + P_i \rightarrow pyruvate + NADPH + H^+ + ATP$$

On the other hand, the serine pathway consumes 2ATP per pyruvate formed. Determinations of molar growth yields reflect these factors, e.g. values for growth on methanol of bacteria using the hexulose phosphate pathway were some 17 to 44% higher than for organisms using the serine pathway.

The yeast dihydroxyacetone pathway consumes 1 ATP. In contrast, there is a very high energy requirement when CO_2 is used for net assimilation rather than formaldehyde.

The energetics of methylotrophic growth are of paramount importance in the commercial exploitation of these organisms for single-cell protein production. An important characteristic for productivity and profitability

Table 9.1 Energy balances for bacterial and yeast C_1-assimilation pathways normalized to pyruvate formation.

Cycle	Route	Reactants	Energy change		
			ΔNAD(P)H	ΔFPH$_2$	ΔATP
Calvin (ribulose bisphosphate)		$3CO_2$	-5	0	-7
Hexulose phosphate (ribulose monophosphate)	FBP aldolase/sedoheptulose bisphosphatase	3HCHO	$+1$	0	0
	FBP aldolase/transaldolase	3HCHO	$+1$	0	$+1$
	KDPG aldolase/sedoheptulose bisphosphatase	3HCHO	$+1$	0	-3
	KDPG aldolase/transaldolase	3HCHO	$+1$	0	0
Serine pathway		$2HCHO + CO_2$	-2	$+1$	-2
Dihydroxyacetone (yeast)		3HCHO	$+1$	0	-1

Abbreviations: FBP, fructose 1, 6-bisphosphate; KDPG, 2-oxo-3-deoxy-6-phosphogluconate; FP, flavoprotein cf succinate dehydrogenase.
Adapted from Quayle and Ferenci (1978), *Microbiol. Revs.* **42**, 251–273.

is a high growth yield. The organism used by ICI Ltd for the production of 'Pruteen' from methanol is a genetically-engineered strain of *Methylophilus methylotrophus*, an organism which employs the hexulose phosphate cycle for the oxidation of formaldehyde produced by methanol dehydrogenase.

CHAPTER TEN

ENERGY CONSERVATION IN BACTERIAL PHOTOSYNTHESIS

10.1 General characteristics of bacterial photosynthesis

In photosynthesis radiant energy is trapped by means of membrane-bound, light-harvesting pigments and used by photoautotrophs, such as green plants, algae and cyanobacteria, to oxidize water and reduce CO_2, producing carbohydrate and releasing molecular oxygen in the overall reaction

$$2H_2O + CO_2 \xrightarrow{\text{light}} (CH_2O) + H_2O + O_2$$

which is referred to as oxygenic photosynthesis. However, the majority of photoautotrophic bacteria carry out anoxygenic photosynthesis (without releasing oxygen), in which water is replaced by other reductants such as S^{2-}, $S_2O_3^{2-}$ or H_2. There are also some photoheterotrophic bacteria which replace both water and CO_2 by partially reduced organic compounds such as malate.

The energy derived from light is used to generate ATP (i.e. photophosphorylation) via a proton circuit which is analogous to that discussed for mitochondrial and bacterial membranes (Chapter 5). Thus a protonmotive force generated across a proton-impermeable membrane is used to drive a proton-translocating ATP synthetase.

The overall process of photosynthesis can be dissected into a sequence of four well-defined phases: (1) radiant energy capture by light-harvesting or antenna pigments (chlorophyll, bacteriochlorophyll and carotenoids); (2) energy transfer to reaction centres containing specialized photopigments—(bacterio)chlorophyll and (bacterio)pheophytin—where charge separation occurs and the energy is converted into the concentrations of chemical components which can be expressed as an electrochemical (redox) potential; (3) electron transfer and proton-coupled

133

electron transfer reactions via a chain of redox carriers comprising quinones, cytochromes and iron-sulphur proteins; and (4) conversion of the energy derived from electron transfer into forms which the cell can use for ATP synthesis, solute transport, reductive assimilation and biosynthesis.

10.2 The photosynthetic bacteria

Microbial photosynthesis is confined to six bacterial families: Rhodospirillaceae, Chromatiaceae, Chlorobiaceae, Chloroflexaceae, Halobacteriaceae and Cyanobacteriaceae. As already noted, only the cyanobacteria carry out oxygenic photosynthesis. In contrast, photosynthesis by the Rhodospirillaceae, Chromatiaceae, Chlorobiaceae and Chloroflexaceae is an anaerobic, anoxygenic process with sulphur compounds, hydrogen or organic substrates serving as reductants, while the Halobacteriaceae are obligate aerobes possessing an entirely different type of photosynthetic apparatus from the rest. The characteristics of these bacteria will now be surveyed as a prelude to a discussion of their energy-transducing mechanisms.

The Chromatiaceae, or purple sulphur bacteria, and the Chlorobiaceae, or green sulphur bacteria, are obligate anaerobes that utilize reduced sulphur compounds such as hydrogen sulphide and thiosulphate as hydrogen donors. The green and purple sulphur bacteria are primarily autotrophic and photoassimilate only simple compounds like acetate and butyrate.

The Rhodospirillaceae or purple non-sulphur bacteria are principally photoheterotrophs although they can also reduce CO_2; unlike most photosynthetic bacteria they are able to grow chemoheterotrophically in the presence of oxygen and thus can multiply in the dark, deriving energy from the oxidation of organic compounds and using oxygen as the terminal electron acceptor. Species from all three genera are able to utilize H_2 as reductant for CO_2.

The Chloroflexaceae or green gliding bacteria resemble the purple non-sulphur bacteria in being primarily heterotrophic and growing well aerobically in the dark.

10.3 The photosynthetic apparatus and its mechanism

There are differences in the arrangement of the photosynthetic apparatus of these organisms. That of the green sulphur bacteria is arranged in apparently discrete vesicles enclosed by an electron-dense membrane; these organelles lie immediately beneath the cytoplasmic membrane. The photosynthetic apparatus of the purple organisms is located in an intra-cytoplasmic membrane formed by deep invaginations of the cytoplasmic membrane (Drews and Oelze, 1981).

The halobacteria are obligate chemoheterotrophic aerobes adapted to life in the very high concentrations of NaCl and Mg^{2+} found in shallow salt lakes (about 5M NaCl is optimum for their growth) where the intensity of illumination is usually high. They are unique among the photosynthetic bacteria in possessing bacteriorhodopsin, a purple-coloured protein, as photopigment; it is located in the cytoplasmic membrane.

Among these various organisms three types of bacterial photosynthesis can be recognized on the basis of the photopigment employed, namely (1) chlorophyll (cyanobacteria), (2) bacteriochlorophyll (green and purple bacteria), and (3) bacteriorhodopsin (halobacteria). Common to the membranes of all of these organisms except the halobacteria are photosynthetic units, each comprising a large number of *antenna* pigment molecules which harvest photons and associated with a single *photochemical reaction centre* to which the trapped light energy is transferred (Glazer, 1983); an electron transfer chain completes the complex (Dutton and Prince, 1978). The size of the unit generally varies in response to environmental changes in order to optimize light trapping, e.g. there may be from 50 to 500 antenna bacteriochlorophyll molecules per reaction centre in purple bacteria. There are two types of reaction centre in oxygenic organisms (one is responsible for the photolysis of water), linked by a single electron–transfer chain, but only one centre in anoxygenic bacteria.

The initial event in photosynthesis is the absorption of a photon by a (bacterio)chlorophyll molecule which is thereby transformed into a singlet excited state, i.e. one electron is raised to a higher energy level. The excited molecule can act as a reducing agent at a much lower redox potential than in the ground state and thus reduces a suitable acceptor, e.g. the energized electron is ejected at a potential about one V more negative. The photochemically harvested energy is represented by the difference in redox potential between the reduced acceptor and oxidized (bacterio)chlorophyll molecule. Unlike the (bacterio)chlorophyll of the reaction centre, excited antenna (bacterio)chlorophyll molecules do not undergo photo-oxidation and their harvested energy is directed by non-radiative inductive resonance transfer between adjacent molecules to the reaction centre where it is trapped. Although some energy may be lost as heat or as fluorescence in the transfer process, the (bacterio)chlorophyll in the reaction centre resides in an environment which decreases the energy needed for its excitation relative to that required for the antenna molecules; energy transfer is thus unidirectional from the antenna pigments to the reaction centre.

The subsequent oxidation of the chlorophyll or bacteriochlorophyll molecule in the reaction centre then occurs but the fate of the ejected electron depends upon the system. Bacterial photosynthesis differs from that of chloroplasts because the electron enters an electron transfer pathway and is eventually returned to the reaction centre with concomitant use of the redox potential to pump protons, i.e. it is a *cyclic* process. In chloroplasts the electron transfer pathway is *non-cyclic* and, as already noted, involves a second reaction centre. One centre (PS_{II}), on excitation, extracts electrons from water ($H_2O \rightleftharpoons 2H^+ + 2e^- + \frac{1}{2}O_2$) and transfers them via a proton-translocating electron transfer chain to a second reaction centre, PS_I, from whence they reduce $NADP^+$ at a redox potential some 1.1V more negative than that of the initiating oxidation of water. The Δp generated is employed for ATP synthesis. Because the bacterial system is cyclic it cannot generate the reducing power needed for CO_2 assimilation and electrons must therefore be derived from the appropriate reductant ($H_2S, S_2O_3^{2-}, H_2$ etc.).

The most intensive research has been conducted with purple bacteria, and our present knowledge of the green sulphur bacteria with respect to details of their reaction centres, electron transfer chains and ATP synthesis is much more fragmentary.

10.4 The photopigments and associated carriers

Present in the membrane-bound photosynthetic units of purple and green bacteria are the photopigments bacteriochlorophyll (BChl), bacteriopheophytin (BPh) and carotenoids, and they are associated with various redox carriers, including quinones, iron-sulphur proteins and cytochromes. The antenna bacteriochlorophyll is usually characteristic of the bacterial species and, together with the light-harvesting carotene (which serves principally to protect the cell from photochemical damage, its energy-transfer role being minor), confers the typical colour on the organisms as well as defining the spectrum of radiation which can be harvested. Reaction centres in bacteria use light of longer wavelength and lower energy than cyanobacteria, algae and higher plants.

Bacteriochlorophyll is a porphyrin molecule with a centrally-bound magnesium (Mg^{2+}) of which five closely-related types are known, designated BChl *a, b, c, d* and *e* respectively. For each bacteriochlorophyll there is a corresponding bacteriopheophytin in which the Mg^{2+} is replaced by $2H^+$.

The reaction centres of the various organisms display some variation but that of a typical purple bacterium contains two dimers of bacteriochlorophyll a and 2 bacteriopheophytins in association with 2 quinones, 1 iron atom and other redox carriers. A considerable amount of research has been done on the isolation and purification of antenna photopigments and reaction centres from some of the Rhodospirillaceae and Chromatiaceae, and a clearer understanding of these systems is emerging.

10.5 The light reaction and energy transduction in the purple bacteria

The initial photochemical event in the reaction centre of a purple bacterium occurs when a dimer of bacteriochlorophyll $(BChl)_2$, designated P_{870} (where P represents 'specialized pigment' and the subscript refers to the wavelength of the major absorption band) undergoes excitation to P^*_{870}. Charge separation leads to electron transfer via bacteriopheophytin (BPh) to one of the two bound ubiquinones (Q_I), giving an anionic semiquinone Q_I^-; the Fe^{2+} atom of the reaction centre is associated with the two quinone molecules but does not undergo a redox cycle and its role is still unclear. These reactions are very rapid, which enables them to compete effectively with energy dissipation via fluorescence, e.g. electron transfer from P^*_{870} to BPh takes about 10 ps and an additional 150 ps for transfer to the ubiquinone. The E_m of P_{870} in its ground state is approximately $+450 \, mV$ and after excitation about $-550 \, mV$, i.e. 1V more negative, while the BPh/BPh^- couple displays an $E_m \simeq -550 \, mV$.

The pair of ubiquinones now act in a concerted manner as a 'gate' and transduce the single electron photochemical event into a $2e^- + 2H^+$ transfer. This is achieved by electron transfer from the anionic semiquinone Q_I^- to the second ubiquinone, Q_{II}, which is then protonated to $Q_{II}H$; a second photon-generated electron undergoes a similar sequence of events and $Q_{II}H$ becomes fully reduced to $Q_{II}H_2$. Only then are $2e^- + 2H^+$ transferred outside the reaction centre to the electron transfer system via the ubiquinone of the bulk phase of the membrane. These systems involve, additional to the bulk-phase quinone, at least two c-type cytochromes and, according to species, b-type cytochromes and Fe–S proteins. A cytochrome c (often c_2 with $E_m + 300 \, mV$) completes the cycle by donating an electron to P^*_{870} to regenerate P_{870} (Figure 10.1). In some species of purple bacteria this cytochrome c is an integral part of the reaction centre. The details of these electron transfer steps are not clearly defined although it is presumed

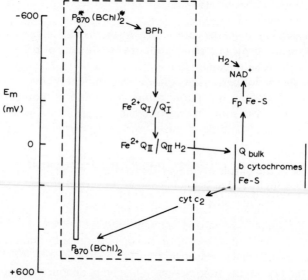

Figure 10.1. Diagrammatic representation of the reaction centre of *Rhodopseudomonas spheroides* and cyclic electron transfer. The broken lines enclose the reaction centre. Abbreviations: P_{870} (BChl)$_2$, the dimer of bacteriochlorophyll with major absorption band at 870nm; Bph, bacteriopheophytin; Q_I, Q_{II}, ubiquinone pair, Fp, flavoprotein.

they resemble those of mitochondria, possibly with a protonmotive Q cycle. Cyclic electron transfer thus generates Δp but cannot, of course, furnish reducing power for biosynthesis; these electrons must be secured from appropriate donors such as reduced sulphur compounds or, in the case of the non-sulphur bacteria, organic substrates. Because the redox components of the cyclic electron transfer sequence all have E_m values more positive than that of the nicotinamide nucleotide couple, NAD^+ cannot be reduced by forward electron transfer and Δp-energized reversed electron flow is required. Here it should be noted that while NAD^+ is the cofactor involved in the purple bacteria, $NADP^+$ operates in the oxygenic organisms. In the former organisms the NADPH needed for biosynthesis is secured via transhydrogenase reactions.

There is evidence that some purple and green bacteria synthesize pyrophosphate as well as ATP during photosynthesis (p. 156).

Measurements of light-induced $\rightarrow H^+/e^-$ quotients have now been made with chromatophores from Rhodospirillaceae, i.e. with inside-out vesicles. Values of 2 were obtained, indicating that the transfer of one electron permits the sequential binding from the external aqueous medium of two protons and their transfer to the interior of the chromatophore, consistent with the mechanism described. Experiments have also demonstrated, and examined the characteristics of, light-induced proton translocation with measurements of Δp, $\Delta\psi$ and ΔpH. Values of Δp in the range of 100 to 420 mV have been recorded with various chromatophores and right-side-out vesicles.

10.6 The light reaction and energy transduction in the green bacteria

The photochemical reactions of the green bacteria are less well understood than those of the purple organisms. However, the antenna bacteriochlorophylls, principally c, d or e, are located in the vesicles while Bchl a and the reaction centres are sited in the adjacent cytoplasmic membrane; it is presumed that some link exists between these membranes. Their carotenoid composition differs from that of purple bacteria. The components of the redox chain comprise at least two c-type cytochromes (*Chlorobium* has three, c_{551}, c_{553}, c_{555}), menaquinone (MK) but not ubiquinones, and several high-potential ($E_m = + 430$ mV) and low-potential ($E_m = - 600$ mV) iron – sulphur proteins, including rubredoxin and ferredoxins.

The primary photochemical event in the reaction centre involves the excitation of the BChl a (P_{840}) raising it from a mid-point potential of $+ 250$ mV to about $- 550$ mV. The electron acceptor is possibly bacteriopheophytin followed by a low potential Fe–S protein. Cyclic electron transport ensues via menaquinone and cytochrome c_{555} ($E_m = + 145$ mV), which reduces P^*_{840} (Figure 10.2). It is noteworthy that the complete cyclic electron transfer system of the green bacteria functions at a significantly lower redox potential than that of the purple organisms. The non-cyclic electron transfer pathway to NAD^+ in the green bacteria appears to possess ferredoxin and flavoproteins and, because of their more negative potentials relative to NAD^+, occurs by forward electron flow. Reducing equivalents from the inorganic sulphur reductants are channelled via c-type cytochromes (Figure 10.2). ATP synthesis occurs at the expense of Δp derived from the electron transfer chain.

10.7 The light reaction and energy transduction in the cyanobacteria

The cyanobacteria carry out oxygenic photosynthesis, similar to that of algae and green plants, employing two photosystems (PS_I and PS_{II}) and non-cyclic electron transfer. Their cytoplasm contains many photosynthe-

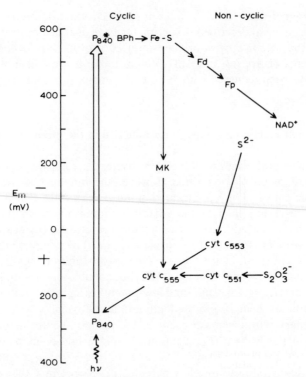

Figure 10.2. Routes of photo-dependent electron transfer in green bacteria. Abbreviations as in Figure 10.1. MK, menaquinone, Fd, ferredoxin.

tic membranes known as *thylakoids*, whose outer surfaces have associated granules termed *phycobilisomes* which possess auxiliary photosynthetic pigments. Most genera of the cyanobacteria are obligatorily photoauto-trophic and many are able to fix atmospheric nitrogen.

The principal photopigment is chlorophyll *a*, and *β*-carotene is the major carotenoid. Cyclic electron transfer can occur in these bacteria via PS_I and involves light harvesting by a bulk chlorophyll *a*–protein complex (B_{680}) with energy transfer to chlorophyll *a* (P_{700}) resident in the reaction centre, possibly via allophycocyanin (a phycobiliprotein) as intermediary. There is strong evidence that the P_{700} of PS_I is a monomer, in contrast to the dimeric form present in the purple and green bacteria. Pheophytin is most likely the primary acceptor for the electron ejected by P^*_{700}, with successive electron transfers to a series of three low-potential Fe–S proteins ($E_m \simeq -530\,mV$). Cyclic electron transfer is completed via *b*-type cytochromes, plas-

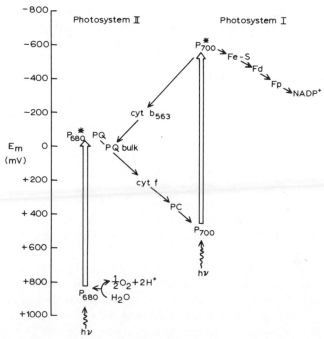

Figure 10.3. Routes of photo-dependent electron transfer in cyanobacteria involving two photosystems and the oxidation of water. Abbreviations: P_{680}, P_{700}, chlorophyll; PQ, plastoquinone; PC, plastocyanin; Fd, ferredoxin; Fp, flavoprotein.

toquinone (PQ, related to ubiquinone, with $E_m \simeq +25\,\text{mV}$), cytochrome f (a c-type cytochrome), and a low molecular weight copper protein, plastocyanin (PC, with $E_m + 370\,\text{mV}$) which donates an electron to P^*_{700} (Figure 10.3).

Photosystem II carries out the photo-oxidation of water and, in association with PS_I, effects non-cyclic electron transfer from water to $NADP^+$. The antenna pigments comprise three phycobiliproteins, namely phycocyanin, phycoerythrin and allophycocyanin, assembled into supramolecular complexes which are the previously noted phycobilisomes. These bodies possess a triangular core of allophycocyanin molecules surrounded by rods of phycocyanin and phycoerythrin; the current belief is that the two peripheral pigments harvest radiant energy, transfer it to allophycocyanin, thence to the bulk chlorophyll B_{680} which, in turn, excites the P_{680} of reaction centre II. B_{680} and P_{680} are both sited in the thylakoid membrane.

Excitation of P_{680} permits electron transfer to the primary plastoquinone acceptor which is reduced to a semiquinone (PQH). Subsequent electron transfer occurs via the bulk quinone pool, a segment of the cyclic pathway and PS_I to the Fe–S centres, and thence to $NADP^+$ by way of ferredoxin and flavoprotein (Figure 10.3).

10.8 The light reaction and energy transduction in the halobacteria: bacteriorhodopsin

When these obligate aerobes grow under a normal atmospheric pressure of oxygen they possess a red cytoplasmic membrane containing carotenoids, which function to protect the cells from photochemical damage, and they employ a conventional cyanide-sensitive respiratory chain. However, when halobacteria grow in the light at the low oxygen concentrations characteristic of their salt lake habitats they synthesize bacteriorhodopsin, a membrane-associated, conjugated protein with a purple retinal (vitamin A aldehyde) prosthetic group. This pigment appears as purple patches in the non-invaginated membrane and may account for as much as 75% of the membrane dry weight, the balance being phospholipid. The pigment is arranged as a single layer in the membrane and each molecule ($M = 26k$) spans the membrane seven times, zig-zag fashion. Such cells display photo-dependent proton extrusion which is insensitive to cyanide. Energy transduction in the halobacteria is therefore achieved either by respiration or photosynthetic electron transfer, in which respect they resemble the Rhodospirillaceae and some of the Cyanobacteriaceae, even though they are devoid of (bacterio)chlorophyll and conventional photosystem redox carriers.

The retinal prosthetic group is covalently linked via a Schiff's base to a lysyl residue of the apo-protein, bacterio-opsin, thus displaying a marked similarity to vertebrate rhodopsin (visual purple). The photochemical reactions of bacteriorhodopsin (bR_{570}) are complex and not fully understood but they involve reversible protonation and deprotonation of the Schiff's base ($- CH = \overset{+}{N}H - \rightleftharpoons - CH = N -$). Investigations using low-temperature and laser-flash spectrometry have revealed the presence of a series of intermediates but, essentially, on illumination, the protonated form (bR_{570}) is bleached, with the release of a proton, to become the bR_{412} form; the membrane orientation of bR is such that the proton is released to the exterior (periplasm). Regeneration of bR_{570} is associated with the uptake of a proton from the cytoplasm: the cycle thus operates as a proton pump by which means one proton is translocated for each photon

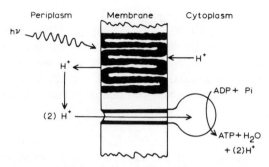

Figure 10.4.   Photosynthetic phosphorylation in *Halobacterium halobium*. The absorption of one photon by the pigment bacteriorhodopsin leads to the translocation of one H^+ across the membrane. The H^+ gradient formed drives the synthesis of ATP via the proton-translocating ATP synthetase (note that $2H^+$ are needed per ATP formed, i.e. 2 photons).

absorbed. The protons return via a proton-translocating ATP synthetase effecting the synthesis of ATP (Figure 10.4).

A model has been proposed which envisages the bacteriorhodopsin encompassing a hydrophilic channel which spans the membrane and ·having a central hydrophobic region which contains the retinal. A conformational change in bR_{570} produced by light is believed to release to the outer channel a proton which is then transferred to the exterior via a sequence of low pK groups such as carboxyls. The deprotonated bR_{412} is

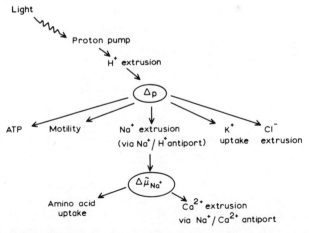

Figure 10.5.   Light-dependent reactions of *Halobacterium halobium*. $\Delta\bar{\mu}_{Na^+}$ is the electrochemical gradient of Na^+ across the membrane. Based on Eisenbach and Caplan (1979).

reconverted to bR_{570} by accepting a proton from the cytoplasm via a sequence of high pK groups, e.g. lysyl, along the inner channel.

The relative ease with which bacteriorhodopsin can be isolated and purified from halobacteria and incorporated into liposomes or model bilayer membranes, has greatly facilitated the direct experimental demonstration of a transmembrane protonmotive force (detected as $\Delta\psi$ and ΔpH) in response to illumination. It is concluded, therefore, that the observed Δp can drive the ATP synthesis required for the uptake of nutrients, growth and motility, and which is also needed to preserve the necessary salt balances that the halophilic life of the bacteria demands. There is, indeed, evidence for the existence of a light-dependent Na^+ pump which generates an electrochemical potential difference of Na^+ ($\Delta\bar{\mu}_{Na}^+$) that can be employed for driving solute uptake (Figure 10.5).

CHAPTER ELEVEN

MICROBIAL ENERGY RESERVE COMPOUNDS

11.1 General considerations

Under certain environmental conditions many micro-organisms are able to accumulate substantial amounts of compounds which can serve as specialized energy reserve materials, akin to the familiar starch and glycogen of plant and animal cells. By furnishing a source of carbon and energy, and in some cases nitrogen, their presence may enable a micro-organism to maintain its viability in the absence of nutrients for a longer period than a corresponding organism which is not so endowed. Although the existence of such materials in microbes was formerly regarded with some scepticism, ample evidence has been obtained to support the concept of microbial energy reserves.

Before an energy storage function can be assigned to any cellular component three criteria should normally be met, namely that (i) the compound is accumulated under conditions when the supply of energy from external sources is in excess of that required by the cell for growth and maintenance at that time; (ii) the compound is utilized when the external supply of energy is no longer adequate for the optimal maintenance of the cell, either for growth and division or for sustaining viability and other processes; and (iii) the compound is degraded to yield energy in a form utilizable by the cell and that it is utilized for some purpose which gives the cell a biological advantage in the struggle for existence over cells which do not possess a comparable compound.

It is clearly undesirable to rely solely on the first two criteria for the evaluation of a storage function, because some compounds might be produced by a cell in an effort to detoxicate end-products of metabolism which would otherwise accumulate at an undesirably rapid rate and prove toxic. If, in criterion (i), emphasis is placed on the *intracellular* accumulation of a compound then a distinction is immediately apparent between storage

145

compounds and 'shunt' or 'overflow' products of metabolism which, for example, are produced in the medium by some fungi.

Three classes of compound have generally been regarded as potential energy storage materials in micro-organisms, namely carbohydrates (polyglucans, glycogen), lipids (including poly-β-hydroxybutyrate) and polyphosphates. Comparatively recently, the existence in cyanobacteria of nitrogen reserve materials (cyanophycin and phycocyanin) has been established. Each of these classes of compound will be considered subsequently, but here we may note certain features common to them all. Thus the cellular content of each of these materials can vary considerably, depending upon the environmental conditions, but since they are of high molecular weight their synthesis and accumulation has relatively little effect on the intracellular osmotic pressure. While some micro-organisms synthesize only one kind of reserve, others are able to accumulate two or more. For example, *Rhodospirillum rubrum* and *Bacillus megaterium* both synthesize glycogen and poly-β-hydroxybutyrate while a *Pseudomonas* species isolated from activated sludge has been reported to accumulate glycogen, poly-β-hydroxybutyrate and lipid. With such organisms the nature of the carbon source used for growth and other environmental factors may then become important in determining the relative amounts of the reserves that are actually accumulated.

11.2 Carbon and energy reserves

11.2.1 Carbohydrates (polyglucans)

Generally micro-organisms which accumulate polyglucans (glycogen and glycogen-like materials) do so under conditions where growth is limited by the supply of utilizable nitrogen and there is an ample supply of the carbon source. However, exceptions are known with both yeast and bacteria, e.g. *Saccharomyces cerevisiae* and *Streptococcus sanguis* have both been shown to synthesize glycogen under carbon-limited conditions in a chemostat. The rate of growth affects the quantity of polyglucan accumulated and an inverse relationship has been observed between growth rate and carbohydrate content with a number of different nitrogen-limited, glucose-grown micro-organisms.

Carbohydrate reserve materials have now been identified and characterized in a wide variety of prokaryotes and thermophilic archaebacteria (Table 11.1). In all these examples the polyglucan is composed of α-D-

Table 11.1 Occurrence of carbohydrate reserve materials in prokaryotes and archaebacteria (the list is not exhaustive)

Prokaryotes	Archaebacteria
Acetivibrio cellulolyticus	*Desulfurococcus mobilis*
Anacystis nidulans	*Desulfurococcus mucosus*
Arthrobacter globiformis	*Sulfolobus acidocaldarius*
Bacteriodes amylogenes	*Sulfolobus solfataricus*
Bacteroides fragilis	*Thermofilum pendens*
Clostridium pasteurianum	*Thermoplasma acidophilum*
Escherichia coli	*Thermoproteus tenax*
Klebsiella aerogenes	
Mycobacterium tuberculosis	
Nocardia asteroides	
Rhodospirillum rubrum	
Ruminococcus albus	
Selenomonas ruminantium	
Streptococcus mutans	
Streptomyces viridochromogenes	
Thiobacillus neapolitans	

glucosyl units linked α-1,4 with α-1,6 branches, and its biosynthesis follows the route shown in Figure 11.1. Following the synthesis of a linear α-1,4 chain of α-D-glucosyl units, a branching enzyme transfers glucosyl units from the non-reducing end of the chain to the 6-position of some units, to produce α-1,6-D-glucosyl linkages.

The first enzyme of the biosynthetic sequence, ADPglucose pyrophosphorylase, is a key regulatory enzyme in most bacteria, being activated by certain glycolytic intermediates (such as fructose 6-phosphate, fructose 1,6-bisphosphate, glycerate 3-phosphate or pyruvate, which presumably serve as signals for a state of 'carbon excess') and inhibited by either ADP, AMP or P_i. The rate of formation of ADPglucose seems thus to be directly related to the adenylate energy charge and maximum velocity to be controlled mainly by a change in the level of activity of ADPglucose pyrophosphorylase. Preiss (1984), who has reviewed bacterial glycogen synthesis, recognized seven groups of ADPglucose pyrophosphorylase enzymes on the basis of the nature of the metabolites activating the enzyme.

Other regulatory factors have also been implicated in an effort to explain how glycogen synthesis is switched on when the nitrogen source is exhausted in the presence of excess of the carbon source, and without an increase in ADPglucose synthase. One scheme for *E. coli* envisages the loss of an inhibitor of glycogen synthesis when exogenous NH_4^+ becomes

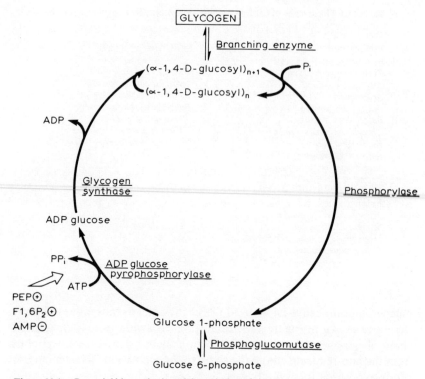

Figure 11.1.   Bacterial biosynthesis and degradation of glycogen and its regulation. As yet, no control of bacterial phosphorylase has been demonstrated.

exhausted. Because NH_4^+ is assimilated by the cell via glutamine it was reasoned that the inhibitor might be an intermediate in a glutamine-dependent pathway, such as *de novo* purine biosynthesis. Evidence has been obtained to support the possible role of 5-aminoimidazole-4-carboxamide ribonucleotide (AICAR), a nitrogen-containing intermediate of purine biosynthesis, as this putative inhibitor of glycogen synthesis; when NH_4^+ is exhausted the intracellular concentration of AICAR would be expected to fall and thus release glycogen synthesis from inhibition.

It has been proposed that the degree of branching of polyglucans might be a factor in determining the rate at which bacteria degrade their reserves. *Escherichia coli* and *Klebsiella aerogenes*, which utilize their polyglucans rapidly, have glycogens which structurally resemble animal glycogen with average chain lengths ($\overline{CL}$) of 12 to 15 glucosyl units, whereas the glycogen of *Pseudomonas* V-10, with $\overline{CL}$ of 8, is structurally similar to those of *Arthrobacter* and *Mycobacterium* ($\overline{CL}$ of 7 to 9) with a high degree of branching, comparable to phosphorylase limit dextrins, which are not susceptible to phosphorylase action unless first debranched by an α-1,6-D-glucosidase enzyme. It is possible, therefore, that the rate of glycogen degradation in these bacteria is limited by the low activity of the debranching enzyme and that this characteristic might contribute to the low rate of glycogen utilization and the long-term survival on starvation observed with both *Arthrobacter* and *Pseudomonas* V-19.

The oral bacterium *Streptococcus sanguis* is an example of an organism which continues to synthesize glycogen under carbon limitation in a chemostat, and an explanation for this behaviour has recently been offered (Keevil *et al.*, 1984). It has been suggested that the regulation of its glycogen synthesis might be at the stage of formation of glucose 1-phosphate rather than of ADPglucose. This bacterium takes up its sugars from the environment by two mechanisms, the phosphoenolpyruvate (PEP)-dependent phosphotransferase system and proton symport (Chapter 6). The intracellular glucose accumulated by the latter route is phosphorylated to glucose 6-phosphate by glucokinase and ATP, or to glucose 1-phosphate using as phosphate donors carbamoyl phosphate (which is derived from arginine metabolism) and/or acetyl phosphate (which is derived from glucose metabolism under conditions when lactate formation is decreased as, for example, under carbon limitation). When *Streptococcus sanguis* is grown at specific growth rates in excess of $0.1 \, h^{-1}$ with glucose limitation, the activities of its PEP-phosphotransferase and glucokinase are inadequate for the phosphorylation of all the available glucose, and thus the other reactions assume prominence, leading to the formation of glucose 1-phosphate and increased glycogen synthesis.

11.2.2 Trehalose

In vegetative cells and spores of fungi the disaccharide trehalose serves as an important storage compound. Trehalose is a non-reducing sugar derived from two molecules of α-D-glucose (linked α-1,1) and generally accumulates in the cell during periods of non-proliferation, whether in the life cycle, the cell cycle or starvation. In all these cases, when growth resumes the stored trehalose is rapidly mobilized. Thus trehalose is accumulated in the reproductive stages of the life cycle and mobilized during germination. In the cell cycle accumulation occurs during the maturation phase and the compound is mobilized prior to cell division. Starvation for specific nutrients in the growth medium, e.g. glucose, nitrogen, phosphate or sulphur in the case of yeast, initiates trehalose synthesis, while replenishment of the medium permits resumption of growth and concomitant trehalose utilization. It is believed that trehalose functions as a storage carbohydrate during the starvation period and the viability of starved yeast has been correlated with its trehalose content. Continuous cultures of yeast grown at low dilution rates also accumulate substantial amounts of trehalose; conversely, fast-growing cultures display a greatly decreased trehalose content.

It has been proposed that trehalose might also furnish energy and intermediary metabolites in yeast spore dormancy, with the intracellular ATP concentration controlling mobilization of the disaccharide. However, trehalose does not appear to serve in this capacity in ascospores of *Neurospora* species which utilize lipids as endogenous substrates during dormancy.

The resistance of fungal spores to extremes of temperature and desiccation seems to be enhanced by a high trehalose content, whereas possession of glycogen does not confer such protection. This observation possibly explains the fact that glycogen is rarely used as a storage material in spores.

The mobilization of trehalose reserves in fungi is apparently regulated by at least two mechanisms. In some fungi modulation of activity of the enzyme trehalase (α, α-trehalose 1-D-glucohydrolase) is believed to occur by a cyclic-AMP-induced phosphorylation of the enzyme. A protein kinase is activated by c-AMP and phosphorylation of the enzyme protein ensues with enhancement of trehalase activity. Other fungi have a non-regulatory trehalase and rapid changes in trehalase activity during trehalose mobilization have never been detected. In these organisms, e.g. *Schizosaccharomyces pombe*, it has been proposed that compartmentation between trehalase and its substrate prevents trehalose hydrolysis. Thus, in *Neurospora* ascospores trehalase is associated with the innermost layer of the cell wall. An appropriate phase transition in the cytoplasmic membrane phospholipids in response to heat or chemical activation would cause a change in permeability of the membrane and permit access of trehalose to the enzyme. Direct evidence for such a mechanism is lacking although the co-existence of high concentrations of trehalose and high trehalase activity in cells is well documented. These regulatory features have been reviewed by Thevelein (1984).

11.3 Polyphosphates

11.3.1 General considerations

The presence of polyphosphate in metachromatically-staining granules of various micro-organisms has been recognized for many years (Kulaev, 1979). It presents a fascinating aspect of microbial energetics, for Fritz Lipmann, in 1965, suggested that the earliest organisms on this planet used polyphosphate or pyrophosphate as their prime energy intermediary, the role of ATP as the universal energy carrier in contemporary organisms having arisen during the course of evolution. This concept receives some support from the fairly recent observation that an increase in the viscosity of the lipid phase of the inner mitochondrial membrane can cause a switch from ATP-generating phosphorylation to pyrophosphate-generating phosphorylation; it is proposed that, during evolution, biomembranes became more fluid, and with this change occurred a transition from phosphorylation yielding pyrophosphate in ancient organisms to phosphorylation with ATP in contemporary ones.

The physiological role of polyphosphate in the microbial cell is, however, not entirely unequivocal and its several functions as an energy reserve compound, phosphorus reserve and regulator of metabolism require careful assessment for individual organisms, although there is no question that polyphosphate can in some micro-organisms fulfil the function of ATP. There have been proposals that polyphosphate possibly represents a metabolic fossil which, over the centuries, has lost its original role in polymer synthesis and has assumed new functions which still elude us; alternatively, it is argued that in the metabolism of contemporary organisms neither polyphosphate nor pyrophosphate has been entirely superseded and they continue to play an important role in metabolism.

The chemical structure of the polyphosphates found in micro-organisms is that of a linear condensed inorganic phosphate (Figure 11.2), varying in chain length from two to about 10^6 units and usually consisting of mixtures of different molecular sizes. Their average chain length is normally

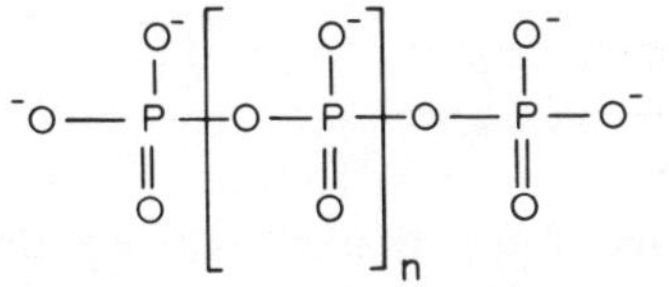

Figure 11.2. Structure of linear inorganic polyphosphate where n may have values from 2 to 10^6.

estimated by titration of the second acid function (with pK_a of about 7) associated with the terminal groups. Thermodynamically polyphosphates are high-energy phosphate compounds and the standard free energy of hydrolysis of the anhydride linkage yields some 38 kJ per phosphate bond at pH 5. Their energy storage function depends on the ability of the bond cleavage reaction to effect phosphorylation and thereby conserve the energy associated with the reaction

$$(P)_n + H_2O \rightarrow (P)_{n-1} + P_i \quad \Delta G^{0\prime} \simeq -38\,kJ\,mole^{-1}$$

where P represents the monomer unit.

Although polyphosphates are not universal cell constituents, their presence has been recorded in bacteria, fungi, algae, mosses, protozoa, insects and in some tissues of higher plants and animals. About 60 diverse species of prokaryotes (bacteria and cyanobacteria) are known to contain polyphosphate granules which generally do not seem to possess a bounding membrane. In cyanobacteria the granules are mostly localized in the region of DNA fibrils and associated ribosomes, and near subcellular structures concerned with photosynthesis. Nuclear magnetic resonance studies with ^{31}P indicate that polyphosphates occur in the periplasmic region of *Mycobacterium smegmatis*, thus resembling findings with eukaryotic micro-organisms. Granules present in *Desulfovibrio gigas* have been rigorously identified as comprising magnesium tripolyphosphate.

In eukaryotes, such as fungi and yeast, a significant proportion of the total polyphosphate is localized on the surface of the cell, either outside the plasma membrane or closely associated with it, and comprising a separate pool from the intracellular polyphosphate. This 'outer' fraction is greater during exponential growth than in the stationary phase. Polyphosphates are absent from mitochondria and also from other structures related to energy generation in eukaryotic protoplasts. The intracellular poly-phosphates of some yeasts are found principally in vacuoles and possibly vesicles of the endoplasmic reticulum, where fractions of different chain length ($n = 5$ and $n = 15$ to 25) have been identified. It seems possible, too, that some of the least polymerized polyphosphates are present in the cytosol of eukaryotes in a free state.

11.3.2 Polyphosphate biosynthesis

Two direct mechanisms of polyphosphate biosynthesis have been identi-fied, neither of which involves pyrophosphate. In some bacteria the transfer of the terminal phosphoryl group of ATP to polyphosphate is catalysed by a Mg^{2+}-dependent ATP-polyphosphate phosphotransferase (polyphosphate kinase), thus

$$(P)_n + ATP \rightarrow (P)_{n+1} + ADP$$

The enzyme is present in diverse aerobic, anaerobic and facultative bacteria and appears to be the unique pathway in *Klebsiella aerogenes* since mutants lacking polyphosphate kinase are unable to synthesize the polymer.

A second biosynthetic system, which is present in *Neurospora crassa* and various bacteria, elongates the polyphosphate chain at the expense of the high energy group potential of 1,3-bisphosphoglycerate, catalysed by a 1,3-bisphosphoglycerate-polyphosphate phosphotransferase

$$
\begin{array}{lll}
CH_2OPO_3^{2-} & & CH_2OPO_3^{2-} \\
| & & | \\
CHOH & + (P)_n \rightarrow & CHOH \qquad + (P)_{n+1} \\
| & & | \\
COOPO_3^{2-} & & CO_2^-
\end{array}
$$

In organisms which possess both enzymes, their relative importance for polyphosphate synthesis is a matter of interest; in *E. coli* maximum activity of the phosphotransferase coincided with the peak period of polyphosphate accumulation whereas the kinase did not display maximum activity until net degradation of the polymer occurred, although the kinase activity did show some increase during the period of maximum polyphosphate deposition.

11.3.3 Polyphosphate degradation

The degradation of polyphosphate in bacteria is catalysed by several enzymes and consequently assessment of their relative contribution to the total rate of polymer breakdown under physiological conditions in the intact cell poses problems. Although the polyphosphate kinase reaction is reversible, being controlled by the ATP:ADP ratio of the cell, such a dual role would not be consistent with the general metabolic observation that biosynthetic and degradative reactions occur via different routes.

Individual enzymes for the transfer of a terminal phosphate from polyphosphate to AMP, glucose and fructose, employing the group transfer potential, have been found in various micro-organisms, as well as polyphosphatases which merely hydrolyse the polymer to inorganic phosphate.

In some Mycobacteria and Corynebacteria a Mg^{2+}-dependent AMP phosphotransferase has been found which catalyses the reaction

$$(P)_n + AMP \rightarrow (P)_{n-1} + ADP$$

but since it has a very high K_m for AMP (20 mM) its physiological significance seems dubious. Of wider distribution and greater importance is a Mg^{2+}-dependent polyphosphate glucokinase, which permits the formation of glucose 6-phosphate from glucose without the intervention of ATP:

$$(P)_n + glucose \rightarrow (P)_{n-1} + glucose\ 6\text{-phosphate}$$

The distribution of this enzyme is of taxonomic interest, being confined to a limited group of organisms (Actinomycetes, Propionibacteria, Micrococci and related species). A similar enzyme, polyphosphate fructokinase, which was discovered in fructose-grown (but not in glucose-grown) *Mycobacterium phlei*, phosphorylates fructose at the expense of polyphosphate:

$$(P)_n + fructose \rightarrow (P)_{n-1} + fructose\ 6\text{-phosphate}$$

Polyphosphatases which hydrolyse long-chain polyphosphates from the terminal groups

$$(P)_n + H_2O \rightarrow (P)_{n-1} + P_i$$

are known. Mutants of *Klebsiella aerogenes* deficient in this enzyme are able to accumulate polyphosphate in the normal way but unable to degrade the polymer. However, the significance of this hydrolytic enzyme in normal metabolism is still uncertain.

The question of whether polyphosphate is degraded to produce pyrophosphate which, as we consider elsewhere (p. 156), can then be used by some micro-organisms in place of ATP, is an open one. Some bacteria isolated from mud by enrichment on tripolyphosphates and tetrapolyphosphates were capable of growth on phosphate compounds, ranging from pyrophosphates to polyphosphates, as energy sources and it has been suggested that conversion to pyrophosphate might be a prelude to utilization of the larger molecules.

11.3.4 Physiological functions of polyphosphate

There is ample evidence available from studies with various microorganisms to support the role of polyphosphate as a reserve of both phosphorus and energy, although the dual function is obviously not characteristic of all organisms investigated. Thus, in *Alcaligenes* species, the polymer is not degraded when the energy supply to the cell is restricted or halted, nor can it maintain the ATP pool under these conditions; however, in this bacterium polyphosphate functions principally as a phosphorus storage compound. Formerly, it was suggested that polyphosphate functioned as a phosphagen, resembling creatine phosphate and arginine phosphate, but this is clearly so only in relation to the provision of phosphate and energy; in all other respects and in their physiological significance these compounds are dissimilar.

The function of polyphosphate as a phosphorus reserve is attested by its ability to supply phosphorus for nucleic acid and phospholipid biosynthesis when micro-organisms are subjected to phosphorus starvation. The properties of the polymer are consistent with this role, for the osmotic equilibrium of the cell would suffer much less change than if inorganic orthophosphate were concentrated, which in turn would be expected to influence the equilibria involving adenine nucleotides and consequently the adenylate energy charge. Bearing in mind the fact that the phosphate content of many natural environments is low due to the insolubility of calcium phosphate, the presence of a phosphorus-reserve material in free-living micro-organisms and the observed derepression of the enzyme responsible for its accumulation under conditions of phosphorus starvation seem eminently reasonable.

The role of polyphosphate in bacterial survival under starvation conditions remains inconclusive. Although two of the essential criteria for designating this polymer as a reserve material can be fulfilled, namely the conditions necessary for its accumulation and utilization, convincing evidence that possession of the polymer aids survival has not yet been forthcoming.

After many years' researches on polyphosphates, Kulaev (1979, 1985) considers the polymer to be a reserve of 'activated phosphate', which plays an important role in microbial metabolism, principally on two counts. First, micro-organisms do not have a homoeostatic system depending on hormonal or nervous regulation and, second, they are very much dependent on environmental conditions, because their cells are in direct contact with the surrounding medium. The possession of polyphosphates is thus seen as a homoeostatic device which enables a micro-organism to accumulate phosphate in a high energy form under conditions of phosphate excess and subsequently to utilize it to permit independence of an unfavourable environment and to enable rapid initiation of growth and development when propitious external conditions are again restored. This concept envisions polyphosphate as the regulator of the cellular concentrations of adenine nucleotides and inorganic phosphate, and thus controlling metabolism via those reactions that are susceptible to the adenylate energy charge.

11.3.5 Role of pyrophosphate in metabolism

During recent years a much greater understanding of the importance of inorganic pyrophosphate in microbial metabolism has been obtained (Reeves, 1976; Wood, 1977). In the presence of Mg^{2+} ions the hydrolysis of pyrophosphate to orthophosphate

$$PP_i + H_2O \rightarrow 2P_i$$

is associated with a rather lower standard free energy of hydrolysis than the corresponding hydrolysis of ATP to ADP, because P_i does not chelate with Mg^{2+} as strongly as does ADP. The values of $\Delta G^{0'}$ (pH 7.0, 38 °C) in the presence of 1 mM free Mg^{2+} (conditions believed to approximate to the situation *in vivo*) are about -31.8 and $-21.7\,\mathrm{kJ\,mol^{-1}}$ respectively. Consequently pyrophosphate has the potential to replace ATP in many biosynthetic reactions. Such examples were first observed with the photosynthetic bacterium *Rhodospirillum rubrum*, in which synthesis of pyrophosphate is coupled to light-induced electron transport. It was shown that the pyrophosphate could serve as an energy source for many reactions, including transhydrogenation, cytochrome reduction and succinate-linked NAD^+ reduction. The enzyme involved, pyrophosphate phosphohydrolase, catalyses the reversible synthesis and hydrolysis of pyrophosphate linked to proton translocation across the chromatophore membrane. It is smaller enzyme ($M \simeq 100\,\mathrm{k}$) than ATP phosphohydrolase (synthetase), specific for inorganic pyrophosphate and Mg^{2+}-dependent. The Δp generated (inside acidic and positive) by pyrophosphate hydrolysis is large enough to drive reversed electron transfer and ATP synthesis, as noted above.

The high-energy function of the anhydride linkage of pyrophosphate can also be utilized in place of ATP in certain other organisms, and very interesting examples are provided by the parasitic protozoan *Entamoeba histolytica* and by *Propionibacterium shermanii*. Both these organisms ferment glucose via the Embden–Meyerhof glycolytic pathway, yet they do not contain significant amounts of either ATP-phosphofructokinase or ATP-pyruvate kinase, the former of these enzymes catalysing a key reaction of this metabolic sequence. It was discovered that they possess instead pyrophosphate fructokinase, catalysing a reaction in which pyrophosphate replaces ATP in the phosphorylation of fructose 6-phosphate to fructose 1, 6-bisphosphate:

$$\text{fructose 6-phosphate} + PP_i \rightarrow \text{fructose 1,6-bisphosphate} + P_i$$

and a pyruvate phosphate dikinase catalysing the reaction

$$\text{phosphoenolpyruvate} + AMP + PP_i \rightleftharpoons \text{pyruvate} + ATP + P_i$$

(It is a dikinase because in the reverse reaction it catalyses the phosphory-lation of both pyruvate and inorganic phosphate.) These organisms also have a carboxytransphosphorylase which carries out a similar reaction to phosphoenolpyruvate carboxykinase except that pyrophosphate re-places ATP:

$$\text{oxaloacetate} + PP_i \rightleftharpoons \text{phosphoenolpyruvate} + CO_2 + P_i$$

They thus differ from animals and some micro-organisms in being able to circumvent the energetically unfavourable synthesis of phosphoenol-pyruvate from pyruvate in a single reaction catalysed by pyruvate phos-phate dikinase, instead of indirectly by coupling the reactions catalysed by pyruvate carboxylase and phosphoenolpyruvate carboxykinase, namely

$$\text{pyruvate} + CO_2 + ATP \rightleftharpoons \text{oxaloacetate} + ADP + P_i$$
$$\text{oxaloacetate} + GTP \rightleftharpoons \text{phosphoenolpyruvate} + GDP + CO_2$$

Sum: $\text{pyruvate} + ATP + GTP \rightleftharpoons \text{phosphoenolpyruvate} + ADP + GDP + P_i$

Two other pyrophosphate-dependent enzymes have been discovered. In *E. histolytica* (but not in *P. shermanii*) there is a pyrophosphate acetate kinase

$$\text{acetate} + PP_i \rightarrow \text{acetyl phosphate} + P_i$$

and in *P. shermanii* (solely) a pyrophosphate serine kinase catalysing the reaction

$$HO.CH_2CH(NH_2).COOH + PP_i \rightarrow H_2O_3PO.CH_2.CH(NH_2)COOH + P_i$$

In the fermentation of glucose by *E. histolytica* the substrate is converted almost quantitatively to ethanol and CO_2, with the proposed ATP yield:

$$2 \text{ glucose} \rightarrow 4 \text{ ethanol} + 4CO_2 + 3ATP$$

It has been suggested that, of the four moles of phosphoenolpyruvate produced from the two moles of glucose, one is metabolized via the pyruvate phosphate dikinase route and three via the carboxytransphos-phorylase pathway. This would have the effect that the $3PP_i$ produced by the latter reaction would provide the $2PP_i$ required for the pyrophos-phate fructokinase reaction and the one needed by the pyruvate phosphate dikinase reaction, and thus balance. Interestingly, this protozoan does not possess a pyrophosphatase and therefore pyrophosphate produced in anabolic reactions would also be available for the two pyrophosphate-dependent reactions of glycolysis.

More recently it has been demonstrated that pyrophosphate can serve as the source of energy for the growth of sulphate-reducing bacteria of the genus *Desulfotomaculum* and other anaerobes in the presence of fixed carbon.

11.4 Polyhydroxyalkanoates

11.4.1 General considerations

Many bacterial species, including Gram-positive and Gram-negative aerobic bacteria, and photosynthetic anaerobic species, lithotrophs and organotrophs, are able to accumulate reserves of the lipid poly-β-hydroxybutyrate (PHB). More recently it has been appreciated that PHB is but one example, albeit the most abundant, of a general class of microbial components termed polyhydroxyalkanoates which have the general formula

$$HO—CH—CH_2—C\left[O—CH—CH_2—C\right]_n O—CH—CH_2—COOH$$

$$\quad\quad\; | \quad\quad\;\; \| \quad\quad\;\; | \quad\quad\;\; \| \quad\quad\; |$$

$$\quad\quad\; R \quad\quad\; O \quad\quad\; R \quad\quad\; O \quad\quad\; R$$

where $R = CH_3$ in the case of PHB.

PHB is thus a linear homopolymer of D($-$)-3-hydroxybutyric acid and is usually of very high molecular weight ($> 10^6$). The molecule is a compact right-handed helix with a twofold screw axis and a fibre repeat period of 5.96Å. It is an ideal reserve material because it is a highly reduced intracellular molecule, and being virtually insoluble exerts negligible osmotic pressure. PHB occurs in granules which are normally spherical and vary in size according to organism; granules with diameters in the range 0.2 to 0.7 μm have been isolated from *Bacillus megaterium*. These granules are enclosed by membranes of between 2.5 and 4.5 nm thickness which have the enzyme PHB polymerase tightly bound to them. The granule membranes of some bacteria, e.g. *Rhodospirillum rubrum* and *Azotobacter beijerinckii*, also possess PHB depolymerase and so these granules are self-hydrolysing, whereas in other organisms the depolymerase is a soluble enzyme.

In 1974 it was first observed that the polymer extracted from bacteria may contain proportions of other 3-hydroxy acids and, with the subsequent application of more refined analytical techniques such as capillary gas-liquid chromatography, the presence of as many as 11 short-chain 3-hydroxy acids has been demonstrated in polymer hydrolysates. For example, the polymer isolated from batch-grown *Bacillus megaterium* contained 95% 3-hydroxybutyrate, 3% 3-hydroxyheptanoate, 2% of an 8-carbon 3-hydroxy acid and trace amounts of 3-hydroxyvalerate, a

6-carbon 3-hydroxy acid and one of unknown chain length. The view has been expressed that all endogenous storage polymers based on 3-hydroxy acids are heteropolymers but so far there is insufficient evidence to support this generalization. Further, it is manifestly untrue of the polymer produced commercially from *Alcaligenes eutrophus* by ICI Ltd under the name of Biopol, which is a homopolymer of 3-hydroxybutyrate.

When the bacterium *Pseudomonas oleovorans* is grown on 50% *n*-octane it accumulates granules which resemble those of PHB but which are composed of poly-β-hydroxyoctanoate. The growth of this organism on a hydrocarbon thus contrasts with that of *Acinetobacter* species which in the presence of hexadecane accumulate the unmodified hydrocarbon.

11.4.2 Factors influencing poly-β-hydroxybutyrate accumulation

The majority of bacteria investigated accumulate PHB in response to a nutrient limitation other than the carbon source. Exceptions have been noted with an asporogenous mutant of *Bacillus megaterium*, with a strain of *Pseudomonas aeruginosa*, and with a *Spirillum* sp., all of which synthesized significant amounts (12 to 18% of the biomass) of the polymer when grown with carbon limitation in a chemostat.

Under appropriate conditions some Azotobacters and *Alcaligenes eutrophus* are able to accumulate up to 80% of their biomass as PHB, and microscopically their cells appear packed with granules. Detailed investigations with glucose-grown *Azotobacter beijerinckii* have shown that only under an oxygen limitation are these very high polymer contents obtained, their concentration being inversely related to the growth rate. Thus cultures grown with nitrogen or carbon limitation rarely contained more than about 3% of their biomass as PHB whereas oxygen-limited cultures accumulated up to 74%, the precise amount depending on the severity of oxygen deprivation and on the imposed dilution (growth) rate in the chemostat. Imposition of an oxygen limitation on a nitrogen-limited culture led to an increased rate of PHB synthesis and polymer accumulation while the converse occurred when an oxygen limitation was relaxed.

Measurements of culture redox potential and intracellular [NADH]/[NAD$^+$] ratios in these experiments supported the concept that PHB synthesis (which is a reductive process) serves as an electron sink for the excess reducing power which accumulates when this obligate aerobe becomes oxygen-limited and electron transport to oxygen via the respiratory chain is curtailed. Glucose metabolism in this organism proceeds by the Entner–Doudoroff pathway with terminal oxidation via the tricarbo-

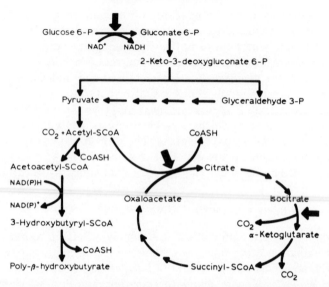

Figure 11.3. Regulation of the Entner–Doudoroff pathway and tricarboxylic acid cycle in *Azotobacter beijerinckii* by NAD(P)H under conditions of oxygen limitation (bold arrows show sites of inhibition). Partial alleviation of the inhibition is afforded by the reductive step in poly-β-hydroxybutyrate formation when acetyl-SCoA is diverted to this pathway because of its restricted entry to the tricarboxylic acid cycle.

xylic acid cycle. Three enzymes in this metabolic sequence, namely glucose 6-phosphate dehydrogenase, citrate synthase and isocitrate dehydrogenase, are powerfully inhibited by either or both of NADH or NADPH, so that when the concentrations of these reduced coenzymes rise as a result of oxygen limitation, glucose metabolism and the operation of the tricarboxylic acid cycle, which generates intermediates and energy for biosynthesis, decrease. The serious consequences of these inhibitory effects for growth of the organism can be largely counteracted by diversion of some acetyl-SCoA from entry to the tricarboxylic acid cycle to PHB synthesis, the reductive step of which (catalysed by acetoacetyl-SCoA reductase) partially alleviates the accumulated reducing power and thereby permits the tricarboxylic acid cycle to operate at a higher rate than would otherwise be possible (Figure 11.3).

11.4.3 The poly-β-hydroxybutyrate metabolic cycle

It was discovered that the biosynthesis and degradation of poly-β-hydroxybutyrate occur via a cyclic process (Figure 11.4). Synthesis of the polymer

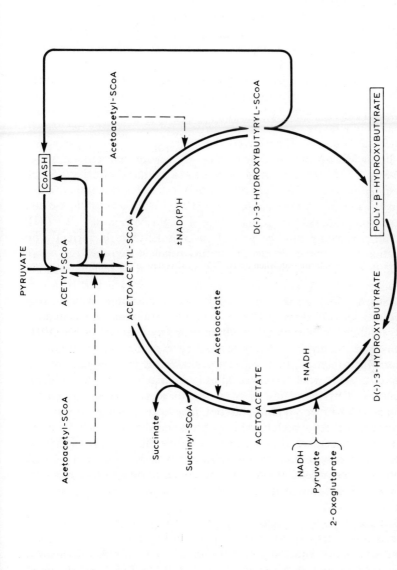

Figure 11.4. Cyclic metabolism of poly-β-hydroxybutyrare and its control in *Azotobacter beijerinckii*. The broken lines indicate the sites of inhibition by the various effectors. After Dawes and Senior (1973). In *Zoogloea ramigera*, which lacks succinyl–SCoA transferase, acetoacetate is converted to its CoA ester by an acetoacetyl-SCoA synthetase.

involves CoA-esters: two molecules of acetyl-SCoA are condensed by the action of β-ketothiolase to release CoASH and form acetoacetyl-SCoA, which is then reduced to D(-)-3-hydroxybutyryl-SCoA, a reaction catalysed by 3-hydroxybutyryl-SCoA dehydrogenase (acetoacetyl-SCoA reductase), an enzyme which utilizes NADPH at about fivefold the rate of NADH. 3-Hydroxybutyryl-SCoA is then the substrate for the granule membrane-bound PHB synthetase (polymerase) which simultaneously liberates CoASH.

Degradation of PHB occurs via a pathway that does not involve CoASH. A depolymerase first releases D(-)-3-hydroxybutyric acid. This enzyme is granule-bound in some bacteria, such as *A. beijerinckii*, but is a soluble enzyme in others. An NAD-specific dehydrogenase oxidizes the acid to acetoacetate which is then converted to acetoacetyl-SCoA by reaction with succinyl-SCoA catalysed by succinyl-SCoA transferase. Thus acetoacetyl-SCoA is an intermediate common to biosynthesis and degradation and regulation is therefore of paramount importance to avoid futile cycling.

Two key regulatory enzymes for the PHB cycle are β-ketothiolase and 3-hydroxybutyrate dehydrogenase; while it might be expected that PHB depolymerase would also be subject to control, there is currently no supporting experimental evidence for this belief. β-Ketothiolase, which has a high K_m (0.9 mM) for acetyl-SCoA, is inhibited in the condensation reaction by free CoASH and in the thiolysis (reverse) reaction by acetoacetyl-SCoA, although the latter inhibition is overcome by increasing concentrations of CoASH. 3-Hydroxybutyrate dehydrogenase is competitively inhibited by NADH, 2-oxoglutarate and pyruvate in *A. beijerinckii*. On this basis, control of the cycle may be explained in the following manner: when there is ample oxygen available the intracellular concentration of CoASH would be expected to be high and that of acetyl-SCoA low, mediated by the action of citrate synthase, with citrate formation serving as a sink for acetyl units and simultaneously releasing free CoASH. The combined effect of high CoASH and low acetyl-SCoA concentrations would prevent acetoacetyl-SCoA formation and thus PHB would not be formed under these conditions. The relief by CoASH of acetoacetyl-SCoA inhibition of acetoacetyl-SCoA thiolysis would, in turn, ensure that only when CoASH was present at high concentrations would PHB degradation proceed.

When the organism becomes oxygen-limited, however, constraint on citrate synthase activity, as a consequence of increased NADH concentration, will occur, and the concentration of acetyl-SCoA will increase with a concomitant decrease in the concentration of CoASH. These conditions thus lead to saturation of β-ketothiolase with acetyl-SCoA and its release from the inhibitory effect of CoASH, so that synthesis of aceto-acetyl-SCoA will occur and PHB synthesis proceed, the reductive step utilizing some of the accumulated reducing power. In the absence of information concerning the control of PHB depolymerase it seems possible that regulation of polymer degradation could be exerted by the inhibition of 3-hydroxybutyrate dehydrogenase by the effectors previously noted. Fine control of PHB metabolism could be exerted by these mechanisms but additonal coarse control also occurs, manifested by changes in the levels of β-ketothiolase and acetoacetyl-SCoA reductase, which increase on oxygen limitation and decrease on its relaxation, mirroring the changes in PHB content of the organism.

Effects of oxygen limitation on other enzymes of *A. beijerinckii* are also observed. NADH oxidase, which plays an important role in the respiratory protection of nitrogenase, decreases and so do isocitrate dehydrogenase and 2-oxoglutarate dehydrogenase. Relaxation of oxygen limitation brings about marked increases in the enzyme levels (Jackson and Dawes, 1976).

The pathways of PHB biosynthesis and degradation seem particularly suited for the role of a redox regulator. The two routes are distinct, reductive biosynthesis utilizing CoA esters and NAD(P)H while oxidative degradation involves the free acids and yields NADH. The polymer serves as an insoluble store of reducing power.

11.4.4 Physiological functions of poly-β-hydroxybutyrate

The polymer has been implicated as a source of carbon and energy for the process of sporulation in *Bacillus* species, although it is not, apparently, a prerequisite for spore formation. A similar conclusion has been reached in relation to its role in the formation of cysts, a process common to the genus *Azotobacter* and which enables organisms to survive adverse conditions.

There is evidence to substantiate the role of PHB as a carbon and/or energy source during starvation of a variety of micro-organisms. Posses-sion of the polymer has been shown to enhance survival under starvation conditions in some, although not all, bacteria studied. One of the most convincing examples of the role of PHB in survival was obtained with a *Spirillum* sp.; resistance to starvation was directly related to its initial PHB content. However, some bacteria with a lower content of PHB seem to survive better than corresponding organisms with a higher one, either because they are not subjected to some additional adverse factor or because they can regulate the utilization of their reserve material more efficiently. A possible additional function of PHB in the Azotobacteriaceae, which are susceptible to inhibition of their nitrogenase by oxygen, is to serve as a substrate for 'oxygen-scavenging' reactions when exogenous substrates are not readily available. Azotobacters display very high respiratory rates and the concomitant utilization of oxygen effectively decreases the partial pressure of oxygen in the immediate environment to tolerable values in the process which has been termed the respiratory protection of nitrogenase. Consequently, the possession of an internal oxidizable substrate such as PHB might be expected to confer advantages over corresponding organisms lacking the polymer when confronted with an inimical oxygen concentration in the absence of external oxidizable compounds.

Poly-β-hydroxybutyrate has also been implicated in the symbiotic nitrogen fixation process between leguminous plants and the bacterial genus *Rhizobium*. These bacteria invade the plant roots, leading to the formation of root nodules in which the micro-organisms exist as transformed pleomorphic cells called *bacteroids* that are no longer capable of independent reproduction. Bacteroids possess an active nitrogenase complex which is not present in free-living *Rhizobium* cells. The significant ATP requirement for nitrogen fixation (p. 56) is met by the metabolism of photosynthetic carbon compounds transported to the root nodules. Evidence for the operation of glycolytic, Entner–Doudoroff, tricarboxylic acid cycle and poly-β-hydroxybutyrate cycle enzymes in rhizobia has been obtained and an interrelationship between nitrogenase activity and PHB accumulation postulated. Thus metabolism of PHB could provide carbon compounds, reducing equivalents via 3-hydroxybutyrate dehydrogenase, and/or ATP in support of nitrogen fixation. Research is continuing in an effort to elucidate the possible interaction between these metabolic processes.

11.5 Cyanophycin and phycocyanin

Micro-organisms do not generally possess reserves of nitrogen, consequently the discovery that cyanobacteria contain two major endogenous nitrogen stores, cyanophycin and an auxiliary photosynthate pigment, phycocyanin, was of considerable interest (reviewed by Allen, 1984). The possibility has been raised that these materials might also serve as energy reserves.

Cyanophycin is found in granules and consists of a branched polypeptide containing arginine and aspartic acid residues in the ratio 1:1 and displaying a molecular weight in the range of 25 000 to 125 000. That from *Anabaena cylindrica* possesses a polyaspartic acid backbone to which arginyl residues are attached by their α-amino groups to each carboxyl group of the polyaspartate backbone, with aspartate occupying both amino- and carboxyl-termini:

$$
\begin{array}{ccc}
\text{Arg} & \left[\begin{array}{c}\text{Arg}\end{array}\right. & \text{Arg} \\
| & | & | \\
\text{NH} & \text{NH} & \text{NH} \\
| & | & | \\
\text{CO} & \text{CO} & \text{CO} \\
| & | & | \\
\text{H}_3\overset{+}{\text{N}} - \text{Asp} & \left.\text{Asp}-\right]_n & \text{Asp}-\text{CO}_2^-
\end{array}
$$

The nature of the peptide bonds in the aspartate core remains to be determined, i.e. whether α-amino to α-carboxyl or to β-carboxyl is involved. The cyanophycin accumulated rarely exceeds about 8% of the biomass. It is synthesized by the action of a synthetase enzyme and degraded by a protease.

Both vegetative cells of *Anabaena* species and their interdependent heterocysts (the sites of nitrogen fixation) contain an exopeptidase which hydrolyses cyanophycin to an aspartate-arginine dipeptide, and the activity of this enzyme, as well as that of the synthetase, is higher in the heterocysts than in vegetative cells. It has been proposed that cyanophycin might therefore play a wider role than that of a nitrogen reserve, functioning as a reservoir between nitrogen fixation and the export of glutamine from heterocysts into vegetative cells, thus separating the constant requirement for fixed nitrogen for biosynthesis from the possibly non-constant rates of nitrogen fixation in the heterocysts.

As *Aphanocapsa* possesses the enzyme arginine dihydrolase, which catalyses the overall reaction

$$\text{arginine} + \text{ADP} + \text{P}_i + \text{H}_2\text{O} \rightarrow \text{ornithine} + \text{CO}_2 + 2\,\text{NH}_3 + \text{ATP}$$

it has been suggested that in these organisms cyanophycin, by furnishing arginine on hydrolysis, can serve as an energy source.

CHAPTER TWELVE

ENERGETICS OF MICROBIAL SURVIVAL

12.1 The nature of the problem

The survival of microbes in the natural environment depends upon their ability to utilize the nutrients that are available to them. These substrates are often of transient appearance and present at very low concentrations. Organisms are therefore likely to experience starvation conditions punctuated by short periods of nutrient supply when relatively rapid growth can occur. Consequently, it is vital that a micro-organism be able to withstand the periods of famine and preserve essential cellular functions which will permit growth to occur when a nutrient supply is restored. Microbes vary considerably in their survival characteristics. Some are endowed with specialized survival mechanisms, such as spore or cyst formation (Sudo and Dworkin, 1973); these dormant forms are much more highly resistant to adverse conditions, including starvation, than the vegetative organisms. However, even vegetative micro-organisms display remarkable differences in their resistance to starvation, according to species and the ecological niche they occupy (Table 12.1). Certain bacteria exist for only a few hours in the absence of nutrients whereas some soil and water organisms can survive for weeks, months or even years.

Preservation of cellular integrity during starvation depends ultimately upon the availability of an energy source and, in the absence of external nutrients (or light in the case of phototrophs), this must be derived from endogenous sources. While some microbes possess specialized reserves of carbon and energy that can be drawn upon in deprivation (Chapter 11), others do not, and catabolism of basal cellular constituents, such as RNA and protein, must provide the maintenance energy needed. Much research effort has been expended in attempting to identify factors critical for survival including utilization of specific cellular components, possession of reserve materials, the endogenous metabolic rate and the adenylate

166

Table 12.1 Survival times of various copiotrophic bacterial species in non-nutrient buffers

Organism	ST_{50} (hours)
Arthrobacter crystallopoietes	2400
Bdellovibrio bacteriovorus	10
Chromatium vinosum	120 (dark)
Escherichia coli	36
Megasphaera elsdenii	
(C limited)	3–5
Methanospirillum hungatii	c. 25
Nocardia corallina	480
Peptococcus prévotii	10–12
Rhodospirillum rubrum	
(light)	344
(dark)	12
Sarcina lutea	65
Selenomonas ruminantium	
(N limited)	3.6–8.1
(C limited)	0.5–2.5
Staphylococcus epidermidis	
(aerobic)	c. 8.0
(anaerobic)	6.0–6.5

ST_{50} is the time required for half the original population to die. The incubation temperatures were in the range 30°–37°C.
(Based on data presented by Mink *et al.*, 1982, and Dawes, 1985, where original references are cited.)

energy charge but in no case has it proved possible to associate death unequivocally with any single parameter (Dawes, 1976). More recently, with a greater understanding of the importance of the energized membrane in transport processes and motility, the role of the membrane potential in survival has been investigated.

12.2 Endogenous metabolism

The endogenous metabolism of a micro-organism is defined as the total metabolic reactions that occur when it is deprived of compounds and elements which may serve specifically as exogenous substrates. These processes must necessarily furnish all the energy needed by non-photosynthetic bacteria for survival under starvation conditions; their relationship with membrane energization is shown schematically in Figure 12.1.

Compounds which have been identified as substrates for endogenous metabolism, in addition to recognised reserve materials, include RNA,

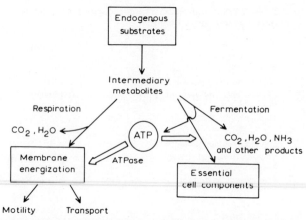

Figure 12.1. Schematic relationship between endogenous metabolism and maintenance energy requirement. In the absence of exogenous substrates, aerobic respiration and anaerobic fermentation of endogenous substrates provide the necessary energy for membrane energization and the resynthesis of essential cell components that may have been degraded during starvation (Dawes, 1985).

protein and, in the case of micrococci, staphylococci and streptococci, certain amino acids of the intracellular free amino-acid pool.

There is now evidence to suggest that the rapid metabolism of endogenous substrates, which generates energy at a rate greatly in excess of the maintenance requirement and which must, therefore, involve energy-spilling reactions, accelerates the death of starved micro-organisms, whereas prolonged survival is associated with a low rate of endogenous metabolism more closely attuned to the maintenance energy requirement. For example, *Escherichia coli*, which in the human gut has to endure relatively short periods of nutrient deprivation, displays a much higher endogenous metabolic rate than does the soil bacterium *Arthrobacter crystallopoietes* which has the capacity to survive many weeks of starvation. However, the concept of energy of maintenance may have little bearing on the special problems posed by the starvation-survival of heterotrophic bacteria in the marine environment. Some of these organisms (termed *oligotrophs* because they can live in extremely nutrient-poor surroundings, as opposed to *copiotrophs* which require much higher concentrations of nutrients for growth) survive for many years in the oceans, most probably by adopting dormant forms characterized by a much smaller cell size, a process which has been termed 'miniaturization'. These small forms, with diameter less than 0.3 μm, are referred to as 'dwarfs' or 'ultramicrobacteria' and display negligible rates of endogenous metabolism. For a survey of this fascinating field of research the reader is referred to the reviews of Morita (1982, 1985).

12.3 The adenylate energy charge and survival

Growing bacteria maintain an adenylate energy charge of 0.8 to 0.9. When growth ceases due to the exhaustion of the carbon and energy source the energy charge falls but, with most organisms investigated, viability is not significantly threatened until the energy charge has dropped to about 0.5; death then intervenes at a rapid rate. However, marked differences of response have been observed with eukaryotes. Thus *Saccharomyces cerevisiae* retained full viability until the energy charge had fallen to 0.3 to 0.1 while *Prototheca zopfii* has been reported to survive carbon starvation for 110 days with an energy charge of 0.01; these differences possibly reflect the more complex situation which exists in the highly compartmented eukaryotic cell. It can be concluded that while energy charge values may be useful predictive indicators of survival for certain microorganisms they are not a universal guide. It must be remembered, too (p. 19), that the adenylate energy charge is a unitless parameter and that, without additional information, it does not provide an indication of either the total size of the adenylate pool or the rate of ATP turnover.

12.4 Storage and energy reserve compounds

Specialized microbial energy reserve and storage compounds are discussed in detail in Chapter 11. The possession by an organism of storage materials like glycogen and poly-β-hydroxybutyrate frequently, but not universally, retards the utilization of cellular components such as RNA and protein during starvation. There is not, however, a common pattern of behaviour and sequential or simultaneous degradation of macromolecules can occur depending upon the organism.

Glycogen has been shown to prolong the viability of a number of bacteria of different genera although in the case of *Escherichia coli* and *Klebsiella aerogenes* it has been suggested that accompanying Mg^{2+} ions might be the principal factor extending survival. Likewise, poly-β-hydroxybutyrate enhances the survival of some though not all micro-organisms investigated (Table 12.2). This polymer also serves as a carbon and energy source for spore formation in some *Bacillus* species but its accumulation is not a prerequisite for the sporulation process. Similarly poly-β-hydro-

Table 12.2 Micro-organisms for which possession of reserve materials has been correlated with enhanced survival on starvation (the list is not exhaustive and the evidence for each organism is not equally compelling).

Glycogen	Poly-β-hydroxybutyrate
Escherichia coli	*Micrococcus halodenitrificans*
Klebsiella aerogenes	*Alcaligenes eutropha*
	Sphaerotilus discophorus
	Azotobacter agilis
	Azotobacter insigne
	Spirillum spp.

xybutyrate has been implicated as a carbon and energy source for the encystment of Azotobacters. However, unequivocal evidence that polyphosphate aids the survival of starved micro-organisms remains to be presented.

During recent years it has been reported that some *Streptococcus* species (*S. lactis, S. faecalis* and *S. cremoris*), when harvested from certain growth media, contain elevated amounts of phosphoenolpyruvate (PEP) and 2-phosphoglycerate which can serve as a source of energy during starvation, supporting sugar transport via the PEP-dependent phosphotransferase system and presumably supplying ATP for maintenance purposes and survival. The increase in intracellular PEP is probably due to inhibition of pyruvate kinase as a consequence of the fall in concentration of early intermediates of glycolysis, e.g. fructose 1,6-bisphosphate, which serve as positive effectors for this enzyme. These streptococci do not possess reserve polymers nor can they use products of RNA degradation as energy sources. Further, since they lack functional cytochromes ATP cannot be generated by oxidative phosphorylation and must be furnished by glycolysis and arginine catabolism. In the early stages of starvation PEP is therefore a valuable source of energy for driving sugar transport.

Addition of small quantities of an appropriate energy source to starving suspensions of bacteria usually prolongs their survival. However, there exists an interesting phenomenon termed *substrate-accelerated death* which is manifest by some organisms whose growth has been limited by the exhaustion of a particular carbon substrate; when that specific carbon source is replenished in a non-nutrient buffered suspension the cells die rapidly. Protection against substrate-accelerated death is afforded by Mg^{2+}, but since these ions are present in many environments the phenomenon is probably not of great significance for an organism in its natural habitat.

12.5 Role of the membrane potential

Survival clearly depends upon the ability of a microbe to take up nutrients, often at very low external concentrations, when they become available in the environment, i.e. upon the preservation of functional transport systems which permit accumulation of substrates against a concentration gradient. As we have discussed elsewhere (Chapter 6), these transport processes are energy-dependent and the maintenance of an energized membrane is crucial for their operation. Obligate anaerobes, or facultative organisms under anaerobic conditions, provide useful model systems for the investigation of the energetics of survival because they energize their membranes with ATP (derived from substrate-level phosphorylation) directly by membrane-bound ATPases. ATP depletion, with consequent loss of transport activity, would effectively prevent the uptake of nutrients at a rate sufficient to support growth when such substrates become available again. Experiments with *Staphylococcus epidermidis* subjected to starvation in phosphate buffer demonstrated a decline in transport activity for glucose and serine on starvation. The membrane potential component ($\Delta\psi$) of the protonmotive force, which is the major driving force for serine uptake in this organism, was also found to decay in concert with loss of viability in the majority of experiments. The generation of membrane potential during starvation could be stimulated by pulses of glucose and the decline in viability correspondingly retarded; as the loss in transport activity and fall in energy charge somewhat preceded the drop in viability, it may be that these parameters have to fall below certain threshold values before an organism is unable to recover from the stress induced by starvation (Horan *et al.*, 1981). However, the results are not entirely unequivocal and further research is needed and with other bacteria. Further, the capacity for survival in an environment which lacks only one essential nutrient that an organism needs for growth is likely to be much greater than in a medium such as phosphate buffer, which represents a very severe deprivation of nutrients.

12.6 Variable proton–solute stoichiometry and 'gating'

Some bacteria possess mechanisms which enable them to retard the rate of energy or metabolite pool dissipation. These comprise symport systems with variable proton-solute stoichiometry, i.e. the number of protons they translocate per solute molecule varies in response to environmental conditions (Konings and Booth, 1981). This device enables an organism to maintain a high rate of solute uptake and a high intracellular concentration of metabolites and ions by means of increasing the proton–solute stoichiometry as the protonmotive force decreases; it is obviously a mechanism of importance to the organism during starvation for it favours the uptake of available nutrients while simultaneously preventing the loss of metabolites from the cell.

An important discovery was that when the protonmotive force falls below a critical value, inhibition of certain transport and energy-transducing systems occurs, a phenomenon referred to as 'gating' (Lanyi and Silverman, 1979). Amino-acid transport in *Halobacterium halobium* and the ATPase of *Streptococcus lactis* were the first systems shown to be affected in this way. The consequence of gating is that ATP is not hydrolysed by the ATPase, Δp is unable to effect carrier-mediated solute uptake, and intracellular metabolites cannot exit via carriers. Under these circumstances metabolite pools can be lost to the cell only via passive diffusion, and this retardation of the loss of energy sources should, for example, enhance survival under starvation conditions.

12.7 The growth precursor cell

Budding and prosthecate micro-organisms that undergo complex morphogenetic cell cycles give rise to cells which differ from their parents or siblings in initiating a cell cycle only in response to an environmental stimulus. These so-called growth precursor cells possess various attributes which fit them for a role as survival cells, among them low endogenous metabolic activity, the absence of rRNA and DNA synthesis and the ability to monitor the environment to determine when differentiation is propitious (Dow *et al.*, 1983). The nature of the environmental stimuli remain to be elucidated, as does the possibility that an organism such as *Escherichia coli* growing at long division times (as would be expected in the natural environment) might also exhibit some form of growth precursor cell.

FURTHER READING

Chapter 1

Arnold, W. N. (1981) *Yeast Cell Envelopes: Biochemistry, Biophysics and Ultrastructure*. Vols. I and II. Boca Raton: CRC Press.

Cabib, E., Roberts, R. and Bowers, B. (1982) Synthesis of the yeast cell wall and its regulation. *Ann. Rev. Biochem.* **51**, 763–793.

Ellar, D. J. (1978) Membrane fluidity in microorganisms. In *Companion to Microbiology*, eds. A. T. Bull and P. M. Meadow, London: Longman, pp. 265–295.

Harrison, R. and Lunt, G. G. (1980) *Biological Membranes: their Structure and Function*. 2nd edn. Glasgow and London: Blackie.

Konings, W. N. (1977) Active transport of solutes in bacterial membrane vesicles. *Adv. Microbial Physiol.* **15**, 175–251.

Lehninger, A. L. (1982) *Principles of Biochemistry*. New York: Worth.

Nikaido, H. and Vaari, M. (1985) Molecular basis of bacterial outer membrane permeability. *Microbiol. Revs.* **49**, 1–32.

Prebble, J. N. (1981) *Mitochondria, Chloroplasts and Bacterial Membranes*. London: Longman.

Rogers, H. J. (1983) *Bacterial Cell Structure*. Wokingham: Van Nostrand Reinhold (UK).

Rogers, H. J., Perkins, H. R. and Ward, J. B. (1980) *Microbial Cell Walls and Membranes*. London: Chapman and Hall.

Salton, M. J. R. and Owen, P. (1976) Bacterial membrane structure. *Ann. Rev. Microbiol.* **30**, 451–482.

Singer, S. J. and Nicolson, G. L. (1972) The fluid mosaic model of the structure of cell membranes. *Science* **175**, 720–731.

Stanier, R. Y., Adelberg, E. A. and Ingraham, J. L. (1977) *General Microbiology*. 4th edn. Englewood Cliffs, N. J.: Prentice Hall.

Williams, R. J. P. (1978) The multifarious couplings of energy transduction. *Biochim. Biophys. Acta* **505**, 1–44.

Chapter 2

Atkinson, D. E. (1968) The energy charge of the adenylate pool as a regulatory parameter. Interaction with feedback modifiers. *Biochemistry* **7**, 4030–4034.

Atkinson, D. E. (1977) *Cellular Energy Metabolism and its Regulation*. New York: Academic Press.

Blair, J. McD. (1970) Magnesium, potassium and the adenylate kinase equilibrium. Magnesium as a feedback signal from the adenine nucleotide pool. *Eur. J. Biochem.* **13**, 384–390.

Bomsel, J. L. and Pradet, A. (1968) Study of adenosine 5'-mono, di and triphosphates in plant tissues. IV. Regulation of the level of nucleotides, *in vivo*, by adenylate kinase: theoretical and experimental study. *Biochim. Biophys. Acta* **162**, 230–242.

Dawes, E. A. (1980) *Quantitative Problems in Biochemistry.* 6th edn. London: Longman, Chapter 3.
Knowles, C. J. (1977) Microbial metabolic regulation by adenine nucleotide pools. In *Microbial Energetics*, eds. B. A. Haddock and W. A. Hamilton, Cambridge University Press, pp. 241–283.
Lehninger, A. L. (1982) *Principles of Biochemistry.* New York: Worth, Chapter 14.
Reeves, R. E. (1976) How useful is the energy in inorganic pyrophosphate? *Trends in Biochemical Sciences* 1, 53–55.

Chapter 3

Dagley, S. (1978) Pathways for the utilization of organic growth substrates. In *The Bacteria, Vol. VI*, eds. I. C. Gunsalus, L. N. Ornston and J. R. Sokatch, New York and London: Academic Press, pp. 305–388.
Dawes, E. A. and Large, P. J. (1982) Class I reactions: supply of carbon skeletons. In *Biochemistry of Bacterial Growth*, eds. J. Mandelstam, K. McQuillen and I. Dawes, Oxford: Blackwell, pp. 125–158.
Dawes, E. A., Midgley, M. and Whiting, P. M. (1976). Control of transport systems for glucose, gluconate and 2-oxogluconate, and of glucose metabolism in *Pseudomonas aeruginosa*. In *Continuous Culture 6: Applications and New Fields*, eds. A. C. R. Dean, D. C. Ellwood, C. G. T. Evans and J. Melling, Chichester: Ellis Horwood, pp. 195–207.
Duine, J. A. and Frank, J. (1981a) Quinoprotein alcohol dehydrogenase from a non-methylotroph *Acinetobacter calcoaceticus*. *J. Gen. Microbiol.* 122, 201–209.
Duine, J. A. and Frank, J. (1981b) Quinoproteins: a novel class of dehydrogenases. *Trends in Biochemical Sciences* 6, 278–280.
Gest, H. (1981) Evolution of the citric acid cycle and respiratory energy conversion in prokaryotes. *FEMS Microbiol. Letts.* 12, 209–215.
Lessie, T. G. and Phibbs, P. V. Jr (1984) Alternative pathways of carbohydrate utilization in Pseudomonads. *Ann. Rev. Microbiol.* 38, 359–387.
Nimmo, H. G. (1984) Control of *Escherichia coli* isocitrate dehydrogenase: an example of protein phosphorylation in a prokaryote. *Trends in Biochemical Sciences* 9, 475–478.
Thauer, R. K. (1982) Dissimilatory sulphate reduction with acetate as electron donor. *Phil. Trans. Roy. Soc. Lond.* B298, 467–471.
Thauer, R. K., Jungermann, K. and Decker, K. (1977) Energy conservation in chemotrophic anaerobic bacteria. *Bacteriol. Revs.* 41, 100–180.
Weitzman, P. D. J. (1981) Unity and diversity in some bacterial citric acid-cycle enzymes. *Adv. Microbial Physiol.* 21, 185–244.

Chapter 4

Bauchop, T. and Elsden, S. R. (1960) The growth of micro-organisms in relation to their energy supply. *J. Gen. Microbiol.* 23, 457–469.
Harrison, D. E. F. (1978) Efficiency of microbial growth. In *Companion to Microbiology*, eds. A. T. Bull and P. M. Meadow, London: Longman, pp. 155–179.
Monod, J. (1942) *Recherches sur la Croissance des Cultures Bactériennes.* Paris: Hermann et Cie.
Pirt, S. J. (1965) The maintenance energy of bacteria in growing cultures. *Proc. Roy. Soc. Lond.,* B 163, 224–231.
Pirt, S. J. (1975) *Principles of Microbe and Cell Cultivation.* Oxford: Blackwell.
Pirt, S. J. (1982) Maintenance energy: a general model for energy-limited and energy-sufficient growth. *Arch. Microbiol.* 133, 300–302.

Stouthamer, A. H. (1976) *Yield Studies in Micro-organisms.* Shildon: Meadowfield Press.
Stouthamer, A. H. (1977) Energetic aspects of the growth of micro-organisms. In *Microbial Energetics*, eds. B. A. Haddock and W. A. Hamilton, Cambridge University Press, pp. 285–315.
Stouthamer, A. H. (1978) Energy-yielding pathways. In *The Bacteria* Vol. VI, eds. I. C. Gunsalus, L. N. Ornston and J. R. Sokatch, New York and London: Academic Press, pp. 389–462.
Stouthamer, A. H. (1979) The search for correlation between theoretical and experimental growth yields. In *International Reviews of Biochemistry. Microbial Biochemistry*, ed. J. R. Quayle, Baltimore: University Park Press, pp. 1–47.
Tempest, D. W. (1978) The biochemical significance of microbial growth yields: a reassessment. *Trends in Biochemical Sciences* **3**, 180–184.
Tempest, D. W. and Neijssel, O. M. (1984) The status of Y_{ATP} and maintenance energy as biologically interpretable phenomena. *Ann. Rev. Microbiol.* **38**, 459–486.

Chapter 5

Amzel, L. M. and Pedersen, P. L. (1983) Proton ATPases: structure and mechanism. *Ann. Rev. Biochem.* **52**, 801–824.
Bashford, C. L. and Smith, J. C. (1979) The use of optical probes to monitor membrane potential. *Meth. Enzymol.* **55**, 569–586.
Boyer, P. D. (1977) Conformational coupling in oxidative phosphorylation and photophosphorylation. *Trends in Biochemical Sciences* **2**, 38–41.
Cross, R. L. (1981) The mechanism and regulation of ATP synthesis by F_1-ATPases. *Ann. Rev. Biochem.* **50**, 681–714.
Ferguson, S. J. and Sorgato, M. C. (1982). Proton electrochemical gradients and energy-transduction processes. *Ann. Rev. Biochem.* **51**, 185–217.
Fillingame, R. H. (1980) The proton-translocating pumps of oxidative phosphorylation. *Ann. Rev. Biochem.* **49**, 1079–1113.
Fillingame, R. H. (1981) Biochemistry and genetics of bacterial H^+-translocating ATPases. *Current Topics in Bioenergetics* **11**, 35–106.
Futai, M. and Kanazawa, H. (1983) Structure and function of proton-translocating adenosine triphosphatase (F_0F_1): biochemical and molecular biological approaches. *Microbiol. Revs.* **47**, 285–312.
Gibson, F. (1982) The biochemical and genetic approach to the study of bioenergetics with the use of *Escherichia coli*: progress and prospects. *Proc. Roy. Soc. Lond.* **B215**, 1–18.
Harold, F. M. (1978) Vectorial metabolism. In *The Bacteria*, Vol. VI, eds. I. C. Gunsalus, L. N. Ornston and J. R. Sokatch, New York and London: Academic Press, pp. 463–521.
Jones, C. W. (1981) *Biological Energy Conservation: Oxidative Phosphorylation.* 2nd edn. London and New York: Chapman and Hall.
Klingenberg, M. (1981) Membrane protein oligomeric structure and transport function. *Nature* **290**, 449–454.
Maloney, P. C. (1982) Energy coupling to ATP synthesis by the proton-translocating ATPase. *J. Membr. Biol.* **67**, 1–12.
Mitchell, P. (1966) Chemiosmotic coupling in oxidative and photosynthetic phosphorylation. *Biol. Revs.* **41**, 445–502.
Nicolay, K., Kaptein, R., Hellingwerf, K. J. and Konings, W. N. (1981) ^{31}P Nuclear magnetic resonance studies of energy transduction in *Rhodopseudomonas sphaeroides*. *Eur. J. Biochem.* **116**, 191–197.
Prebble, J. N. (1981) *Mitochondria, Chloroplasts and Bacterial Membranes.* London: Longman.
Ramos, S., Schuldiner, S. and Kaback, H. R. (1979) The use of flow dialysis for determinations of ΔpH and active transport. *Meth. Enzymol.* **55**, 680–688.

176 MICROBIAL ENERGETICS

Rogers, H. J. (1983) *Bacterial Cell Structure*. Wokingham: Van Nostrand Reinhold (UK).
Rottenberg, H. (1979) The measurement of membrane potential and ΔpH in cells, organelles, and vesicles. *Meth. Enzymol.* 55, 547–569.
Senior, A. E. and Wise, J. G. (1983) The proton ATPase of bacteria and mitochondria. *J. Membr. Biol.* 73, 105–124.
Simon, M., Silverman, M., Matsumura, P., Ridgway, H., Komeda, Y. and Hilmen, M. (1978) Structure and function of bacterial flagella. In *Relations between Structure and Function in the Prokaryotic Cell*, eds. R. Y. Stanier, H. J. Rogers and B. J. Ward, Cambridge University Press, pp. 271–284.
Skulachev, V. P. (1979) Membrane potential and reconstitution. *Meth. Enzymol.* 55, 586–603.
Skulachev, V. P. and Hinkle, P. C. (1981) *Chemiosmotic Proton Circuits in Biological Membranes*. Reading, Mass.: Addison Wesley.
Smith, D. G. (1978) Bacterial motility and chemotaxis. In *Companion to Microbiology*, eds. A. T. Bull and P. M. Meadow, London: Longman, pp. 321–341.
Taylor, B. L. (1983a) Role of proton motive force in sensory transduction in bacteria. *Ann. Rev. Microbiol.* 37, 551–573.
Taylor, B. L. (1983b) How do bacteria find the optimal concentration of oxygen? *Trends in Biochemical Sciences* 8, 438–441.
Ugurbil, K., Shulman, R. G. and Brown, T. R. (1979) High-resolution ^{31}P and ^{13}C nuclear magnetic resonance studies of *Escherichia coli* cells *in vivo*. In *Biological Applications of Magnetic Resonance*, ed. R. G. Shulman, New York and London: Academic Press, pp. 537–589.
Ziegler, M. M. and Baldwin, T. O. (1981) Biochemistry of bacterial bioluminescence. *Current Topics in Bioenergetics* 12, 65–113.

Chapter 6

Anthony, C. (1980) Methanol as substrate: theoretical aspects. In *Hydrocarbons in Biotechnology*, eds. D. E. F. Harrison, I. J. Higgins and R. Watkinson, London: Heyden, pp. 35–57.
Boulton, C. A. and Ratledge, C. (1984) The physiology of hydrocarbon-utilizing microorganisms. *Topics in Enzyme and Fermentation Biotechnology* 9, 11–77.
Dawes, E. A. (1980) *Quantitative Problems in Biochemistry*. 6th edn. London: Longman, Chapter 11.
Eddy, A. A. (1982) Mechanisms of solute transport in selected eukaryotic micro-organisms. *Adv. Microbial Physiol.* 23, 1–78.
Hamilton, W. A. (1975) Energy coupling in microbial transport. *Adv. Microbial Physiol.* 12, 1–53.
Harold, F. M. (1978) Vectorial metabolism. In *The Bacteria* Vol. VI, eds. I. C. Gunsalus, L. N. Ornston and J. R. Sokatch, New York: Academic Press, pp. 463–521.
Harold, F. M. (1982) Pumps and currents: a biological perspective. *Current Topics in Membranes and Transport* 16, 485–516.
Konings, W. N. (1977) Active transport of solutes in bacterial membrane vesicles. *Adv. Microbial Physiol.* 15, 175–251.
Konings, W. N. and Veldkamp, H. (1983) Energy transduction and solute transport mechanisms in relation to environments occupied by micro-organisms. In *Microbes in their Natural Environments*, eds. J. H. Slater, R. Whittenbury and J. W. T. Wimpenny, Cambridge University Press, pp. 153–186.
Maloney, P. C. (1982) Coupling between H^+ entry and ATP synthesis in bacteria. *Current Topics in Membranes and Transport* 16, 175–193.
Prebble, J. N. (1981) *Mitochondria, Chloroplasts and Bacterial Membranes*. London: Longman.
Scarborough, G. A. (1985) The mechanisms of energization of solute transport in fungi. In *Environmental Regulation of Microbial Metabolism*, eds. I. S. Kulaev, E. A. Dawes and D. W.

Tempest, New York: Academic Press, pp. 39–51.
Tanford, C. (1983) Mechanism of free energy coupling in active transport. *Ann. Rev. Biochem.* **52**, 379–409.

Chapter 7

Haddock, B. A. and Jones, C. W. (1977) Bacterial respiration. *Bacteriol. Revs.* **41**, 47–99.
Ingledew, W. J. and Poole, R. K. (1984) The respiratory chains of *Escherichia coli*. *Microbiol. Revs.* **48**, 222–271.
John, P. and Whatley, F. R. (1975) *Paracoccus denitrificans* and the evolutionary origin of the mitochondrion. *Nature* **254**, 495–8.
John, P. and Whatley, F. R. (1977) The bioenergetics of *Paracoccus denitrificans*. *Biochim. Biophys. Acta* **463**, 129–153.
Jones, C. W. (1977) Aerobic respiratory systems in bacteria. In *Microbial Energetics*, eds. B. A. Haddock and W. A. Hamilton, Cambridge University Press, pp. 23–59.
Jones, C. W. (1979) Energy metabolism in aerobes. In *International Reviews of Biochemistry*. *Microbial Biochemistry*, ed. J. R. Quayle, Baltimore: University Park Press, pp. 49–84.
Jones, C. W. (1982) *Bacterial Respiration and Photosynthesis*. Wokingham: Van Nostrand Reinhold (UK).
Jurtshuk, P. and Yang, T.-Y. (1980) Oxygen reactive haemoprotein components in bacterial respiratory systems. In *Diversity of Bacterial Respiratory Systems* Vol. 1, ed. C. J. Knowles, Boca Raton, Florida: CRC Press, pp. 137–159.
Lehninger, A. L. (1975) *Biochemistry*. 2nd edn. New York: Worth.
Lloyd, D. and Turner, G. (1980) Structure, function, biogenesis and genetics of mitochondria. In *The Eukaryotic Microbial Cell*, ed. G. W. Gooday, D. Lloyd and A. J. P. Trinci, Cambridge University Press, pp. 143–180.
Mitchell, P. (1975) The protonmotive Q cycle: a general formulation. *FEBS Letters* **59**, 137–139.
Mitchell, R. A. (1984) Enzyme-catalysed oxygen exchange reactions and their implications for energy coupling. *Current Topics in Bioenergetics* **13**, 203–255.
Nicholls, D. G. (1982) *Bioenergetics: An Introduction to the Chemiosmotic Theory*. London and New York: Academic Press.

Chapter 8

Gottschalk, G. and Andreesen, J. R. (1979) Energy metabolism in anaerobes. In *International Reviews of Biochemistry*. *Microbial Biochemistry*, ed. J. R. Quayle, Baltimore: University Park Press, pp. 85–115.
Kröger, A. (1977) Phosphorylative electron transport with fumarate and nitrate as terminal hydrogen acceptors. In *Microbial Energetics*, eds. B. A. Haddock and W. A. Hamilton, Cambridge University Press, pp. 61–93.
Kröger, A. (1978) Fumarate as terminal acceptor of phosphorylative electron transport. *Biochim. Biophys. Acta* **505**, 129–145.
Odom, J. M. and Peck, H. D. Jr (1984) Hydrogenase, electron transfer proteins, and energy coupling in the sulfate-reducing bacteria *Desulfovibrio*. *Ann. Rev. Microbiol.* **38**, 551–592.
Peck, H. D. Jr (1984) Physiological diversity of the sulfate-reducing bacteria. In *Microbial Chemoautotrophy*, eds. W. R. Strohl and O. H. Tuovinen, Columbus: Ohio State University Press, pp. 309–335.
Postgate, J. R. (1984) *The Sulphate-Reducing Bacteria*. 2nd edn. Cambridge University Press.
Thauer, R. K. Jungermann, K. and Decker, K. (1977) Energy conservation in chemotrophic anaerobic bacteria. *Bacteriol. Revs.* **41**, 100–180.

Vignais, P. M., Henry, P-M., Sim, E. and Kell, D. B. (1981) The electron transport system and hydrogenase of *Paracoccus denitrificans*. *Current Topics in Bioenergetics* 12, 115–196.

Chapter 9

Aleem, M. I. H. (1977) Coupling of energy with electron transfer reactions in chemolitho trophic bacteria. In *Microbial Energetics*, eds. B. A. Haddock and W. A. Hamilton, Cambridge University Press, pp. 351–381.

Anthony, C. (1982) *The Biochemistry of Methylotrophs*. London and New York: Academic Press.

Cobley, J. G. (1976) Reduction of cytochromes by nitrite in electron-transport particles from *Nitrobacter winogradskyi*. Proposal of a mechanism for H^+ translocation. *Biochem. J.* 156, 493–498.

Cobley, J. G. and Cox, J. C. (1983). Energy conservation in acidophilic bacteria. *Microbiol. Revs.* 47, 579–595.

Cox, J. C. and Brand, M. D. (1984) Iron oxidation and energy conservation in chemoauto-trophs: the thermodynamics of ATP synthesis in the acidophile *Thiobacillus ferro-oxidans*. In *Microbial Chemoautotrophy*, eds. W. R. Strohl and O. H. Tuovinen, Columbus: Ohio State University Press. pp. 31–46.

Ingledew, W. J. (1982) *Thiobacillus ferrooxidans*: the bioenergetics of an acidophilic chemolithotroph. *Biochim. Biophys. Acta* 683, 89–117.

Ingledew, W. J., Cox, J. C. and Halling, P. J. (1977) A proposed mechanism for energy conservation during Fe^{2+} oxidation by *Thiobacillus ferro-oxidans*: chemiosmotic coupling to net H^+ influx. *FEMS Microbiol. Letts.* 2, 193–197.

Jones, C. W. (1982) *Bacterial Respiration and Photosynthesis*. Wokingham: Van Nostrand Reinhold (UK).

Kelly, D. P. (1978) Bioenergetics of chemolithotrophic bacteria. In *Companion to Micro-biology*, eds. A. T. Bull and P. M. Meadow, London and New York: Longman, pp. 363–386.

Kelly, D. P. (1985) Physiology of the thiobacilli: elucidating the sulphur oxidation pathway. *Microbiol. Sci.* 2, 105–122.

Large, P. J. (1983) *Methylotrophy and Methanogenesis*. Wokingham: Van Nostrand Reinhold (UK).

Schlegel, H. G. (1975) Mechanisms of chemo-autotrophy. In *Marine Ecology*, Vol. II, ed. O. Kinne, London: Wiley, pp. 9–60.

Strohl, W. R. and Tuovinen, O. H. eds. (1984) *Microbial Chemoautotrophy*. Columbus: Ohio State University Press.

Veenhuis, M., Van Dijken, J. P. and Harder, W. (1983) The significance of peroxisomes in the metabolism of one-carbon compounds in yeasts. *Adv. Microbial Physiol.* 24, 1–82.

Whittenbury, R. and Kelly, D. P. (1977) Autotrophy: a conceptual phoenix. In *Microbial Energetics*, eds. B. A. Haddock and W. A. Hamilton, Cambridge University Press, pp. 121–149.

Zeikus, J. G. (1983) Metabolism of one-carbon compounds by chemotrophic anaerobes. *Adv. Microbial Physiol.* 24, 215–299.

Chapter 10

Baccarini-Melandri, A., Casadio, R. and Melandri, B. A. (1981) Electron transfer, proton translocation, and ATP synthesis in bacterial chromatophores. *Current Topics in Bioenergetics* 12, 197–258.

Clayton, R. K. and Sistrom, W. R. (1978) *The Photosynthetic Bacteria*. New York and London: Plenum Press.

Drews, G. and Oelze, J. (1981) Organization and differentiation of membranes of phototrophic bacteria. *Adv. Microbial Physiol.* 21, 1–92.

Dutton, P. L., Mueller, P., O'Keefe, D. P., Packham, N. K., Prince, R. C. and Tiede, D. M. (1982) Electrogenic reactions of the photochemical reaction centre and the ubiquinone-cytochrome b/c_2 oxidoreductase. *Current Topics in Membranes and Transport* **16**, 323–343.

Dutton, P. L. and Prince R. C. (1978) Energy conversion processes in bacterial photosynthesis. In *The Bacteria*, Vol. VI, eds. I. C. Gunsalus, L. N. Ornston and J. R. Sokatch, New York and London: Academic Press, pp. 523–584.

Eisenbach, M. and Caplan, S. R. (1979). The light-driven proton pump of *Halobacterium halobium*; mechanism and function. *Current Topics in Membranes and Transport* **12**, 165–248.

Evans, M. C. W. (1983) Production and dissipation of membrane potential; formation of ATP and reducing equivalents. In *The Phototrophic Bacteria: Anaerobic Life in the Light* ed. J. G. Ormerod, Oxford: Blackwell, pp. 61–75.

Evans, M. C. W. and Heathcote, P. (1983) The early photochemical events in bacterial photsynthesis. In *The Phototrophic Bacteria: Anaerobic Life in the Light*, ed. J. G. Ormerod, Oxford: Blackwell, pp. 35–60.

Glazer, A. N. (1983) Comparative biochemistry of photosynthetic light-harvesting systems. *Ann. Rev. Biochem.* **52**, 125–157.

Govindjee (1983) *Photosynthesis: Energy Conversion by Plants and Bacteria*. Vol. I. New York and London: Academic Press.

Graber, P. (1982) Phosphorylation in chloroplasts: ATP synthesis driven by $\Delta\psi$ and by ΔpH of artificial or light-generated origin. *Current Topics in Membranes and Transport* **16**, 215–245.

Honig, B. (1982) Photochemical charge separation and active transport in the purple membrane. *Current Topics in Membranes and Transport* **16**, 371–382.

Jones, C. W. (1982) *Bacterial Respiration and Photosynthesis*. Wokingham: Van Nostrand Reinhold (UK).

Jones, O. T. G. (1977) Electron transport and ATP synthesis in the photosynthetic bacteria. In *Microbial Energetics*, eds. B. A. Haddock and W. A. Hamilton, Cambridge University Press, pp. 151–183.

Ke, B. (1978) The primary electron acceptors in green plant photosystems I and photosynthetic bacteria. *Current Topics in Bioenergetics* **7**, 75–138.

Kondratieva, E. N. (1979) Interrelation between modes of carbon assimilation and energy production in phototrophic purple and green bacteria. In *International Reviews of Biochemistry. Microbial Biochemistry*, ed. J. R. Quayle, Baltimore: University Park Press, pp. 117–175.

Nugent, J. H. A. (1984) Photosynthetic electron transport in plants and bacteria. *Trends in Biochemical Sciences* **9**, 354–357.

Ormerod, J. G. (ed.) (1983) *The Phototrophic Bacteria: Anaerobic Life in the Light*. Oxford: Blackwell.

Chapter 11

Allen, M. M. (1984) Cyanobacterial cell inclusions. *Ann. Rev. Microbiol.* **38**, 1–25.

Dawes, E. A. (1984) Stress of unbalanced growth and starvation in micro-organisms. In *The Revival of Injured Microbes*, eds. M. H. E. Andrew and A. D. Russell, London: Academic Press, pp. 19–43.

Dawes, E. A. & Senior, P. J. (1973) The role and regulation of energy reserve polymers in micro-organisms. *Adv. Microbial Physiol.* **10**, 135–266.

Jackson, F. A. and Dawes, E. A. (1976) Regulation of the tricarboxylic acid cycle and poly-β-hydroxybutyrate metabolism in *Azotobacter beijerinckii* grown under nitrogen or oxygen limitation. *J. Gen. Microbiol.* **97**, 303–312.

Keevil, C. W., Marsh, P. D. and Ellwood, D. C. (1984) Regulation of glucose metabolism in oral streptococci through independent pathways of glucose 6-phosphate and glucose 1-

phosphate formation. *J. Bacteriol.* **157**, 560–567.

Kulaev, I. S. (1979) *The Biochemistry of Inorganic Polyphosphates*, New York: Wiley.

Kulaev, I. S. (1985) Some aspects of environmental regulation of microbial phosphorus metabolism. In *Environmental Regulation of Microbial Metabolism*, eds. I. S. Kulaev, E. A. Dawes and D. W. Tempest, New York: Academic Press, pp. 1–25.

Kulaev, I. S. and Vagabov, V. M. (1983) Polyphosphate metabolism in micro-organisms. *Adv. Microbial Physiol.* **24**, 83–171.

Merrick, J. M. (1978) Metabolism of reserve materials. In *The Photosynthetic Bacteria*, eds. R. K. Clayton and W. R. Sistrom, New York & London: Plenum Press, pp. 199–219.

Preiss, J. (1984) Bacterial glycogen synthesis and its regulation. *Ann. Rev. Microbiol.* **38**, 419–458.

Reeves, R. E. (1976) How useful is the energy in inorganic pyrophosphate? *Trends in Biochemical Sciences* **1**, 53–55.

Thevelein, J. M. (1984) Regulation of trehalose mobilization in fungi. *Microbiol. Revs.* **48**, 42–59.

Wood, H. G. (1977) Some reactions in which inorganic pyrophosphate replaces ATP and serves as a source of energy. *Fed. Proc.* **36**, 2197–2205.

Chapter 12

Dawes, E. A. (1976) Endogenous metabolism and the survival of starved prokaryotes. In *The Survival of Vegetative Microbes*, eds. T. R. G. Gray and J. R. Postgate, Cambridge University Press, pp. 19–53.

Dawes, E. A. (1985) Starvation, survival and energy reserves. In *Oligotrophic and Copiotrophic Bacteria in Natural Environments*, ed. M. Fletcher, London: Academic Press.

Dow, C. S., Whittenbury, R. and Carr, N. G. (1983) The 'shut down' or 'growth precursor' cell – an adaptation for survival in a potentially hostile environment. In *Microbes in their Natural Environments*, eds. J. W. Slater, R. Whittenbury and J. W. T. Wimpenny, Cambridge University Press, pp. 187–247.

Gray, T. R. G. and Postgate, J. R. (eds.) (1976) *The Survival of Vegetative Microbes*. London: Cambridge University Press.

Horan, N. J., Midgley, M. and Dawes, E. A. (1981) Effect of starvation on transport, membrane potential and survival of *Staphylococcus epidermidis* under anaerobic conditions. *J. Gen. Microbiol.* **127**, 223–230.

Konings, W. N. and Booth, I. R. (1981) Do the stoichiometries of ion-linked transport systems vary? *Trends in Biochemical Sciences* **6**, 257–262.

Lanyi, J. K. and Silverman, M. P. (1979) Gating effects in *Halobacterium halobium* membrane transport. *J. Biol. Chem.* **254**, 4750–4755.

Mink, R. W., Patterson, J. A. and Hespell, R. B. (1982) Changes in viability, cell composition, and enzyme levels during starvation of continuously cultured (ammonia-limited) *Selenomonas ruminantium*. *Appl. Env. Microbiol.* **44**, 913–922.

Morita, R. Y. (1982) Starvation-survival of heterotrophs in the marine environment. *Adv. Microbial Ecol.* **6**, 171–198.

Morita, R. Y. (1985) Starvation and miniaturisation of heterotrophs, with special emphasis on maintenance of the starved viable state. In *Oligotrophic and Copiotrophic Bacteria in Natural Environments*, ed. M. Fletcher London: Academic Press.

Sudo, S. Z. and Dworkin, M. (1973) Comparative biology of prokaryotic resting cells. *Adv. Microbial Physiol.* **6**, 152–224.

Index